BOSTON STUDIES IN THE PHILOSOPHY OF SCIENCE

BOSTON STUDIES
IN THE PHILOSOPHY
OF SCIENCE

PROCEEDINGS OF THE BOSTON COLLOQUIUM

FOR THE PHILOSOPHY OF SCIENCE

1961/1962

Edited by

MARX W. WARTOFSKY

Department of Philosophy, Boston University,

Boston, Mass., U.S.A.

D. REIDEL PUBLISHING COMPANY

DORDRECHT-HOLLAND

SYNTHESE LIBRARY

A SERIES OF MONOGRAPHS ON THE

RECENT DEVELOPMENT OF SYMBOLIC LOGIC,

SIGNIFICS, SOCIOLOGY OF LANGUAGE,

SOCIOLOGY OF SCIENCE AND OF KNOWLEDGE,

STATISTICS OF LANGUAGE

AND RELATED FIELDS

Editors:

B. H. KAZEMIER/D. VUYSJE

1963
Softcover reprint of the hardcover 1st Edition 1963

ISBN-13: 978-94-010-3265-0   e-ISBN-13: 978-94-010-3263-6
DOI: 10.1007/978-94-010-3263-6

CONTENTS

CONTENTS

PREFACE

The broad range of interdisciplinary concerns which are encompassed by the philosophy of science have this much in common: (1) they arise from reflection upon the fundamental concepts, the formal structures, and the methodology of the sciences; (2) they touch upon the characteristically philosophical questions of ontology and epistemology in a unique way, bringing to traditional conceptions the analytic apparatus of modern logic, and the new content and conceptual models of active scientific investigations.

These sources are reflected in the present volume, which consists of the major portion of the papers presented to the Boston Colloquium for the Philosophy of Science in the academic year 1961–1962. There is no central theme nor any dominant approach in this colloquium. Initiated in 1960 as an inter-university interdisciplinary faculty group, the Colloquium is intended to foster creative and regular exchange of research and opinion, to provide a forum for professional discussion in the philosophy of science, and to stimulate the development of academic programs in philosophy of science in the colleges and universities of metropolitan Boston. The base of the Colloquium is our philosophic and scientific community, as broad and heterodox as the academic, cultural and technological complex in and about this city. The Colloquium has been supported in its first full year, as an inter-institutional cooperative association, by a generous grant to Boston University from the U.S. National Science Foundation. We are most grateful for this help. Many of these papers were published in issues of *Synthese*; they are collected here, with additional materials, in a single volume. We hope it will be found useful and interesting.

We are grateful to Professor Philipp Frank, and to Professors George D. W. Berry of Boston University, Gerald J. Holton and Israel Scheffler of Harvard University, Frederick Schick of Brandeis University, and William A. Wallace, O.P., of the Dominican House of Studies at Dover, for their advice. We wish also to thank Richard M. Millard, then acting Dean, and the Boston University Graduate School for supporting the first half-year of the Colloquium in Spring 1961, and to thank the present

PREFACE

Dean, Richard S. Bear, for his continuing interest. Particular thanks are
due to our secretary, Joan Ringelheim, and to Victor Van Neste, Jr., who
tape-recorded the meetings and helped to transcribe the proceedings.
And, of course, the heart of the matter: we wish to thank the participants
for making the philosophy of science a matter of lively inquiry.

ROBERT S. COHEN
Chairman of the Colloquium
Department of Physics, Boston University

MARX W. WARTOFSKY
Director of the Colloquium
Department of Philosophy, Boston University

SATOSI WATANABE

A MODEL OF MIND-BODY RELATION IN
TERMS OF MODULAR LOGIC

Presented October 26, 1961

PART I

PHILOSOPHICAL BACKGROUND AND POSITION OF PROBLEM

1. *Introduction*

This paper is intended to show that if we assume a certain set of postulates about the mind-body relation, we are forced to face a paradox of logical nature, and that this paradox can be resolved by introducing a rigorously formalizable modification of a certain axiom of the Boolean logic [1]. While I admit that my argument for this set of postulates is by no means conclusive, the paradox thus exposed seems to underlie, in different and more ill-defined forms, many of the confused discussions in the past about the so-called mind-body puzzle. Due to the nature of our approach, we shall be largely concerned with the linguo-logical aspect of the mind-body problem and hence the metaphysical aspect of the mind-body problem has relevance to this work only to the extent that any sensible metaphysical argument must first take a clear stand about this logical paradox.[2] In Part I, I shall lay down the postulates which lead to the paradox, assuming a position as if I were entirely convinced of the unavoidability of these postulates, and in Part II, I shall show the way to resolve this paradox. Relegating a discussion in terms of a more careful language to later Sections, let us describe here, in a loose manner, the nature of the problem we are facing. The first requirement for any honest discussion of the problem is to take full cognizance of the fact that there are prima facie two kinds of phenomena, mental and physical, and correspondingly two kinds of parlances. The problem is to clarify the relationship between these two 'aspects'. Any theoretical view which is based on a method which from the beginning forces us to close our eyes and ears on one side of the two 'aspects' and forbids us to talk about it, is bound to be incapable to solve the problem because there can be no problem in this view. In other words, behavioristic trisimianism, even though useful as an approximate method in certain scientific research, disqualifies as a philosophical method to cope with broader problems because it is not sufficiently 'comprehensive'. By comprehensive is meant here a property of taking into account all pertinent aspects which are prima facie given.

In this way, it may seem that we can *formally* eliminate a behavioristic view from the arena, but if behaviorism comes back with the allies of neurophysiology and physics, (let us call this allied army physicalism), it is not so easy to repulse them, although the same blame of not being sufficiently comprehensive can be placed on them. The reason is that they can now claim that their description of a person in terms of behavior and neurophysiology and physics is 'complete'. If the maximal description of all electrons and nucleons in a person (such as the quantum state) is given, then what else more is there to be added to the picture? Isn't a quantum state, by definition, a description which does not allow any further specification?

To this argument, my first answer will be that a complete description and a comprehensive description are not the same thing. By a complete description of an object is meant one in which every detail of the object 'is given a place' in it, but this does not necessarily mean that every detail 'is given a right place in a right perspective'. By way of a loose metaphor, suppose a horizontal plan of a house is given. This plan is complete in the sense that every small detail of the house is marked somewhere in the plan; in fact nothing is missing. But this plan is not comprehensive because the plan combined with the elevation will reveal the mutual relations of the details which were not revealed by the plan alone. A complete description is the most detailed description perceived from a certain angle and contains every detail, yet it does not tell the whole story. In the same way, a physicalistic description of a person is a complete description, yet it can be inadequate.

This answer of mine, however, does not give any solution; it only poses a question. Resorting again to our loose metaphor, we have to ask how we can stereoscopically synthesize the physical description and the mental description. One might think that this is an easy task; all we need is to do exactly what we should do in the presence of a plan and an elevation of a

2

house. In this case, we just make a logical conjunction of a statement that a given point P has such and such position in the plan and another statement that the same point P has such and such height according to the elevation, and the logical conjunction places the point P in the right perspective.[3] But is this prescription feasible in the case of the physical plan and the mental elevation? In the first place, there can be a viewpoint, more or less in line with Ryle's famous contention [1], that a conjunction of a mental and physical statement just does not make any sense, on the ground that it involves a 'category mistake'. If we take this view, we have to abandon the entire synthesizing effort, and go back to either one of the two kinds of description. This is neither a solution nor a dissolution of the problem, but just a return to a trisimianism. On the other hand, if we take a view that such a hybrid conjunction has a meaning, then we get into another dilemma. Either we have to concede that a physically maximal description (quantum state) can be further subdivided into more detailed descriptions by adding a mental coordinate, or else we shall have to contend that by the use of the mental coordinate, nothing more is actually added. The first alternative is hardly acceptable, for it interferes with the completeness of physical description. The second alternative is hardly a solution since it implies that the entire mental parlance is (extensionally) redundant. This impasse is essentially what I shall call the fundamental paradox of the mind-body problem[4], and I believe that although it may not be pronounced so explicitly elsewhere as is done in this paper, this paradox underlies the entire mind-body controversy and it is not honest not to face it squarely. The solution I give in this paper is such that the synthesis is possible (i.e., we need not return to a trisimianism) without interfering with the completeness of the physical picture and without condemning the mental parlance as a useless redundancy. This will require a subtle, but rigorously formulatable, modification of the Boolean logic.

In the remainder of this Part I, viz., in Sections 2, 3, 4, and 5, I shall lay down what is believed to be the minimum requirements that any adequate theory of mind-body relation should satisfy, and it will be indicated that these requirements involve a certain paradox. In Part II, this paradox is clearly exposed as a contradiction to the usual logic, and it will be shown that by a modification of the axioms of logic all the main requirements of Chapter I can be satisfied in such a way that the paradox is no longer a paradox. This modified logic will turn out to be nothing novel, but a well-known type of non-Boolean logic which is satisfied by the observational propositions in quantum physics.

The present paper may be characterized as a 'model' theory of mind-body relation, but the term 'model' has to be understood in a double role. First, the present viewpoint is a model theory in the sense that it selects a certain number of conspicuous features of the subject-matter under study and ignores some other details.[5] The above-mentioned 'requirements' determine this set of conspicuous features. Second, it is a model theory in the sense that it introduces at a later stage a geometrical model which actually satisfies all the above-mentioned requirements.[6] This geometrical model is thus a model of the model in the first sense. We need not attach any kind of 'reality' to this geometrical model for it is only a tool to demonstrate the internal consistency of the model in the first sense. In the past, I developed two lines of argument, which are independent of each other and also independent of the one adopted in the present paper, both, however, like the present one, leading to the conclusion that consciousness is irreducible to physical phenomena. The theme of one of them [2] was that 'a robot (computing machine) can have a goal but not a value', where 'goal' is a physical term, while 'value' is a mental term which represents, so to speak, the motor force which keeps consciousness awake.[7] The theme of the second one [3] was that 'the theoretical structure of quantum mechanics is such that there exists no meaningful way of using it, except in an approximate sense, without making reference to the consciousness of an observer'. Few among physicists who have been using quantum mechanics correctly have realized this point, and no philosopher has cared for studying the basic premises of quantum mechanics sufficiently to discover it. Recently, E. P. Wigner [4] made this point very clear. Connections of these two arguments with the one developed in this paper will not be discussed in this paper, partly because such a discussion will obscure the purely logical point of view adopted in this paper, and partly because I have not given sufficient thought to this matter so that I can discuss it publicly.

Finally, at the third annual meeting of the New York University Institute of Philosophy, in May, 1959 [5], I made a brief remark regarding the mind-body duality: 'Either we stick to one or the other of the two 'phases' and keep the distributive law of Boolean logic, or we consider

4

both 'phases' simultaneously and replace the distributive law of logic by a less restrictive law such as a 'modular law', as suggested by Birkoff and von Neumann. In the first case, the description is not comprehensive, and in the latter case, we have to give up the familiar logic. This seems to offer a model on which to build a philosophical dualism.'[8] The present paper may be considered as an elaboration of the idea laid down by these short sentences. Curiously enough, this quotation can still be used as an adequate resumé of the present work. The opinions of other researchers on mind-body problem referred to in the present paper are mainly those expressed in the proceedings of this N.Y.U. meeting. [5]

2. *Postulate of Non-Equivalence*

The equivalence postulate asserts that a certain [9] mental proposition [10] is true if and only if a certain physical proposition is true. This postulate is implied not only by the so-called Identity Theory of Feigl [6], but also by certain epiphenomenalists, dualists, and even by some behaviorists and Gestalt theoreticians. According to deterministic epiphenomenalism, a physical state causes a well-defined mental state. Hence, there must exist a mutual implication of a mental proposition and a certain physical proposition. In a strict (non-interactionistic) parallelistic version of dualism, there exists a detailed one-to-one correspondence between a mental event and a physical event. Behaviorism, in a strict sense of the word, ignores completely the mental propositions, hence there can be no place, by definition, for the equivalence postulate. However, there are philosophers of behavioristic tendency who also adopt the equivalence postulate of some sort. Putnam [7] claims that whatever can be said regarding the relationship between a mental state and the corresponding physical state can be said regarding the relationship between a logical state of a Turing machine and a physical state of the Turing machine. Now, a logical state of a Turing machine takes place if and only if a certain physical state of its hardware realization takes place. The thesis of isomorphism introduced by Gestalt theoreticians tries to establish one-to-one correspondence (the word isomorphism implies among other things the one-to-one correspondence) between a pattern of mental processes and a pattern of brain processes, hence between a mental proposition and a physical proposition. Feigl [6] frees himself from the unproductive discussion as to which of mind and body is the cause and which is the

5

effect and also from the useless redundancy caused by dual substance mind and body, and presents the equivalence postulate in a pure form. He claims certain neurophysiological terms denote (refer to) the very same events (referents) denoted by certain phenomenal terms, where the referents are neither material nor mental but the immediately experienced qualities. I shall now enumerate reasons why I do not give much credibility to the equivalence postulate. They may not be entirely convincing, but they are certainly not less convincing than any argument in support of the postulate of exact equivalence.

(a) *Why two languages for one world?*

More or less for the same reasons that Feigl himself gives, I take his version of the equivalence postulate as the least objectionable. However, Feigl's theory involves a puzzle, if not a dilemma, of its own. One of the advantages of the Feiglian view is that it is 'simplifying', in the sense that 'instead of conceiving two realms or two concomitant types of events, it has only one reality which is represented in two different conceptual systems'. But if one proposition composed in one conceptual system is equivalent to another proposition composed in the other conceptual system, then why should we not push the simplifying effort a step further and reduce the two conceptual systems to a single one? It is often the case that if two concepts are different in intension and identical with respect to the instances which are so far known, then we shall eventually discover an individual instance which belongs to one concept but not to the other, destroying the extensional identity. If their extensions are rigorously identical without exception, then two concepts for one extension are useless redundancy, as far as economy of language is concerned. The reason why we would not be willing to take the last economizing measures may be that we really cannot, due to the fact that a mental term and the apparently corresponding physical term do not have after all an exactly identical referent.[11]

(b) *Can we empirically establish equivalence?*

If a mental proposition A_m and a neurophysiological proposition A_b are claimed to be equivalent, then there must be a simultaneous test of the mental state and of the neurophysiological state which permits us to establish or disprove that truth of A_m is always accompanied by truth of A_b

and vice versa. (This is so because A_m and A_b are not dispositional propositions, but are experiential propositions).[12] But, is such a test possible?[13] It is true that we can take an electroencephalograph record or a lie detector record while we are thinking or talking, and a certain rough probabilistic correlation may be established between the mental process and the neural process. However, in view of the fact that the neural system possesses such a fantastic amplifying capability that even a few photons can produce a drastic change in the state, any description short of a quantum mechanical one is bound to be a probabilistic coarse description which does not designate any definite physical state or any definite set of physical states. As is well known, a quantum mechanical observation is an active one rather than a passive one, in the sense that the resulting description does not refer to the state which the observed system would take in the absence of observation.[14] It is certain that the quantum mechanical observation aiming at the exact description of the neural or cerebral system will not only completely incapacitate the mental faculty but will in all probability kill the person. Hence, the required simultaneous test of the physical and mental states is impossible. This is the kind of argument advanced many years ago by Niels Bohr [8] with regard to the relationship between life phenomena and physical phenomena.

(c) *A mental proposition corresponds to a dispositional, physical proposition*

Ryle [1] contends that testing of a mental statement must involve making some hypothetical (dispositional) statement about overt behavior. I agree only partially with this view because I believe that truth of a mental statement regarding my own consciousness can be established under a favorable condition without resorting to observation of my behavior. However, I agree with Ryle on the point that if 'translated' into the behavioral language, a mental statement cannot be made to correspond to an observational proposition whose truth or falsehood can be tested by a single observation. It has to be made to correspond to a probabilistic law-like statement which can never be tested in its entirety, so that we can only evaluate its credibility by observing individual behaviors to which it assigns certain probabilities. This situation invites us to infer that there is no pair of mental and behavioral statements which imply each other. If we add to this argument a plausible contention that there exists a neural statement which implies a given behavioral statement, and vice

7

versa, then we can conclude that an exact equivalence of a mental statement and a neural statement is untenable. I can go a step further, probably against Ryle's intention, and contend that any physical statement, 'translated' into the mental language, becomes a dispositional proposition. In fact, there is no mental proposition which is equivalent to a physical statement, such as 'this rose is red' or 'this electron has its spin in the z-direction'. A physical statement of this sort entails only probabilities to various sense data, which are used as the basis of confirmation of the physical statement. This is analogous in the reverse sense to the case of behavioral confirmation of a mental statement. One may rephrase this situation by saying that for a given mental proposition A_m and a given physical proposition A_b there exist conditional probabilities of the type: $p(A_m \mid A_b)$ and $p(A_b \mid A_m)$ and that these probabilities are never exactly unity.[15] It goes without saying that the existence of a probability does not entail that we can actually evaluate or calculate its value.

(d) *Ordinary language is notoriously interactionistic*

I do not subscribe to the view that the concepts and categories which are entrenched in ordinary language have to be unconditionally upheld, and that one can develop a philosophical view only depending on ordinary language without thoroughly studying the result of ever progressing empirical sciences. However, I do believe that the concepts and categories and the concomitant theoretical frame-work inherent in ordinary language should not be carelessly modified or discarded without a 'solid justification', for they are after all results of a process of adaptation of animal and human mind to experience which extends over many millions of years. What I call a solid justification includes at least the following two conditions: (i) A positive proof must be presented showing that a usual, common-sense concept is utterly incapable of coordinating the pertinent empirical facts or at least that it entails an extreme complexity in coordinating them. (ii) A domain of experience must be clearly designated in which the old, common-sense concept retains its validity and usefulness.[16] These two conditions were amply satisfied when the concept of simultaneity was discarded by Relativity Theory and when the concept of causality was modified by Quantum Theory. If we apply this kind of criteria to the grounds on which the dualistic, interactionistic view entrenched in ordinary language is discarded by some philosophers (among them, ironically, proponents of ordinary language philosophy!), we are surprised to find that their ground of justification is extremely precarious. If oversimplification is permitted for the present, their ground is essentially that we can build a 'consistent' picture of the world without resorting to the interactionistic view. But, possibility of a consistent picture does not prove its adequacy, and in particular, it is bound to be inadequate if a certain aspect or certain facts are deliberately omitted from the picture.[17] Hampshire [9] may be the first to point out that ordinary language is dualistic and Feigl [6] emphasizes that ordinary language is interactionistic. See Ducasse [10] for a valiant, though somewhat weak, defense of interactionism.

Now the interactionist view, coupled with the equivalence postulate, really does not add anything new or alien to the non-interactionist view, for the simple reason that if interactionism claims that a mental event A_m causes a physical event B_b, then non-interactionism can assert that a physical event A_b which is equivalent to A_m causes the physical event B_b. Therefore, if we want to attach to interactionism a validity or an implication which cannot be reproduced by non-interactionism, we simply have to abandon the equivalence postulate.

(e) *Consequence from the postulate of restricted conjoinability*

I shall show later that the equivalence postulate is logically precluded by what I call the principle of restricted conjoinability. My defense of the principle of restricted conjoinability which will presently be given is based on an argument somewhat different from my argument against the equivalence postulate.

As admitted before, the arguments (a) through (e) speaking against the postulate of strict equivalence are rather weak; however, any arguments speaking in support of the postulate are perhaps even weaker. Hence it is at least worth investigating the consequences of the rejection of the postulate. To summarize formally my points developed in this section, I assert the following: (i) *A constant absurdity in the mental parlance and a constant absurdity in the physical parlance are equivalent (denoted hereafter by $\emptyset$). A constant truth in the mental parlance and a constant truth in the physical parlance are equivalent (denoted hereafter by $\square$). (ii) There is no pair (A_m, B_b) of propositions, one mental (A_m) and the other physical (B_b),*

9

which are equivalent to each other, $A_m \leftrightarrows B_b$, or $A_m = B_b$, except in the case where $A_m = \varnothing$ and $B_b = \varnothing$ and the case where $A_m = \square$ and $B_b = \square$. (iii) *We can speak of conditional probabilities of the type* $p(A_m \mid B_b)$ *and* $p(B_b \mid A_m)$, (iv) *which in some cases may become close to unity* (v) *but never becomes exactly unity except in the case where the main event* $[i.e., A_m$ *of* $p(A_m \mid B_b)$ *and* B_b *of* $p(B_b \mid A_m)]$ *is a constant truth.* Point (i) is inevitable since the two parlances refer to the same objects, Point (iv) is a limited admission of the idea of approximate equivalence (which has a certain empirical backing), but point (v) rules out the postulate of strict equivalence, against which I have argued in this sub-section. Point (ii) will hereafter be referred to as 'postulate of non-equivalence'.

Actually, in order to reach the main conclusions of this paper[18], the postulate of non-equivalence as formulated in (ii) above is not necessary (although sufficient) and can be weakened considerably. Namely, the conclusions remain still valid even if for some, but not every, given A_m which is neither $\varnothing$ nor $\square$, there exists B_b such that $A_m = B_b$. For this matter see later Sections. But, for simplicity of argument, we postulate (ii) in the major part of the following discussion.

3. *Atomicity and the Postulate of Restricted Conjoinability, Fundamental Paradox*

Before introducing what I call postulate of restricted conjoinability, I should point out the fact that among physical propositions there are such things as 'maximal description' or 'atomic proposition'.[19] For instance, speaking of a motion of a mass point, we have maximally described its state by specifying its position and its velocity (provided Newton's second law is true). Any more detailed description with an additional specification is either redundant or directly false. Similarly, in quantum mechanics, specification of the 'quantum state' of a system is the minutest description, beyond which any further characterization adds nothing new or leads to absurdity. We thus acknowledge that in the physical parlance, there exist atomic propositions, $\alpha_b, \beta_b, \gamma_b, \ldots$ etc., such that for any physical proposition A_b, either $A_b \cap \alpha_b = \alpha_b$ or $A_b \cap \alpha_b = \varnothing$, where $\varnothing$ denotes a constant absurdity. This can be also restated in the form: $A_b \rightarrow \alpha_b$ implies either $A_b = \varnothing$ or $A_b = \alpha_b$, where the symbol $\rightarrow$ stands for entailment (implication). In connection with atomic propositions, it is often assumed not only that $\alpha_b \cap \beta_b = \varnothing$ and $\alpha_b \cup \beta_b \cup \gamma_b \cup \ldots = \square$, but also that any A_b is a disjunction of some atomic propositions. However, we do not need these properties in our argument here. For symmetry, we shall assume also atomicity of mental propositions later, but this too is not necessary in order to derive the important conclusion of this Section.

Ryle has to be credited for manifestly claiming that it makes no sense to conjoin or disjoin a physical proposition with a mental proposition. See p. 22ff. of reference [1]. I do not think that his analogy in terms of examples such as 'He visited Oxford University and Magdalen College' or 'She came in tears and in a Jaguar' which involve 'category mistakes' is a happy one. I feel that difficulty of a conjunction of a physical proposition and a mental proposition involves a much deeper logical problem than these simple cases of category confusion. But, this is beside the point at present. My main objection is based on the fact that ordinary language is not only constantly making such allegedly non-sensical conjunctions and disjunctions but also conveying some meaningful information through these conjunctions. On the other hand, we have to agree with Ryle that something not quite in order is taking place when such a conjunction is formed – but only odd in the sense that it cannot be located according to the familiar dichotomous frame-work referring to the mental and the physical. However, being neither mental nor physical does not mean being non-sensical. Thus, we may postulate what I call the restricted conjoinability postulate: *The conjunction of a mental proposition ($\neq \square$) with a physical proposition ($\neq \square$) has a meaning (not necessarily a constant absurdity or a constant truth) but is neither a mental nor a physical proposition except in the case where it becomes a constant absurdity. For the case of disjunction, interchange constant absurdity ($\varnothing$) and constant truth ($\square$) in the foregoing sentence.*

It is obvious that the postulate of non-equivalence is a corollary to this conjoinability postulate. Suppose A_m and B_b are respectively a mental proposition and a physical proposition, which are equivalent to each other. It suffices then to show that this premise entails, by the use of the restricted conjoinability postulate, either $A_m = B_b = \varnothing$ or $A_m = B_b = \square$. In fact according to the postulate, A_m and B_b must be (i) $\square$, or (ii) $A_m \cap B_b = \varnothing$ or (iii) $A_m \cap B_b$ must be neither physical nor mental. Since $A_m = B_b$, case (ii) entails $A_m = B_b = \varnothing$, and case (i) entails $A_m = B_b = \square$. Case (iii) is impossible since the relation $A_m = A_m \cap B_b$

11

shows that $A_m \cap B_b$ belongs to the mental parlance. Q.E.D. For later reference, it should be noted that this demonstration uses only laws that are valid in any general lattice and does not use the distributive law.

To make explicit what we have tacitly admitted so far, we add here a postulate which may be called the 'postulate of merger': *The ordinary language* [20] $\mathcal{L}_0$ *is a set of propositions which is closed with respect to conjunction, disjunction and negation. The mental parlance* $\mathcal{L}_m$ *is a subset of* $\mathcal{L}_0$ *which is also closed with respect to the three basic operations. So is the physical parlance* $\mathcal{L}_b$. To simplify our terminology, let us rephrase this postulate by saying that $\mathcal{L}_m$ and $\mathcal{L}_b$ are sub-lattices of the lattice $\mathcal{L}_0$, although a set which is closed with respect to the three operations is not necessarily a lattice, unless something more specific is said about the cardinality of the sets and about the rules (such as idempotent law, associative law, commutative law, distributive law, modular law, law of double negation, etc.) which are supposed to be obeyed by these operations. The postulate of non-equivalence states that the two sub-lattices have only two common elements $\varnothing$ and $\square$.

Now, we are prepared to proceed to expose what I call the fundamental paradox of mind-body relation. This paradox originates essentially from a hidden conflict between the non-equivalence postulate and (a strengthened form of) the atomicity postulate. In a loose usage of words, this conflict reflects the fact that completeness of a view (language) does not warrant its comprehensiveness. In order to see what kind of situation can be expected from the postulate that $\mathcal{L}_m$ and $\mathcal{L}_b$ are sublattices of a larger lattice, let us take examples within the physical parlance. Take as a sublattice $\mathcal{L}_1$ the set of propositions, that can be generated by three atomic propositions α_1: 'The object is blue', β_1: 'The object is yellow', and γ_1: 'The object is red'. The universe of discourse is assumed to be such that any object has to have one of the three colors, and no more detailed color description is possible. Then the lattice $\mathcal{L}_1$ consists of various propositions (their number is 8) that can be formed from these three atomic propositions such as 'the object is not yellow', 'The object is yellow or blue'. Now consider a second sub-lattice $\mathcal{L}_2$ which is generated also by three atomic propositions, α_2, β_2, γ_2. Now, there are two conceivable cases. In one case, there is at least one atom in $\mathcal{L}_1$ which is equivalent to one of the atoms of $\mathcal{L}_2$. In another case, there is not a single atom in $\mathcal{L}_1$ which is equivalent to an atom of $\mathcal{L}_2$. An example of the first case can be constructed by assigning to the atom α_2 of $\mathscr{L}_2$ the following proposition 'the object reflects only the electromagnetic waves in the region of wave-lengths W_B' where W_B corresponds one-to-one to the color blue. Then α_1 and α_2 are equivalent. Let us note that in this case the atom $\alpha_1 = \alpha_2$ remains an atom in the wider lattice which is generated by merger of $\mathscr{L}_1$ and $\mathscr{L}_2$. An example of the second case can be constructed by assigning to α_2, $_{B2}$, γ_2, the propositions: 'The object is large', 'The object is medium-sized', 'The object is small', where the three predicates, large, medium, small are suitably defined. In this case, each atom in $\mathscr{L}_1$ is split into three smaller propositions, which are the atoms of the larger lattice which is formed by merger of $\mathscr{L}_1$ and $\mathscr{L}_2$. For instance, $\alpha_1 \cap \gamma_2$ will stand for 'The object is blue and small' and $\alpha_1 \cap \beta_2$ for 'the object is blue and medium-sized' etc.

Compare now the problem of $\mathscr{L}_m$ and $\mathscr{L}_b$ to this example of $\mathscr{L}_1$ and $\mathscr{L}_2$. The first case where there is an equivalence corresponds to the case where some mental proposition which is neither $\varnothing$ nor $\square$ is equivalent to some physical proposition. This must be excluded on account of the postulate of non-equivalence. But, on the other hand, the second case where an atom of one parlance is further subdivided into smaller propositions by merger with another parlance is against any good common sense. In fact, the premise of this case implies that a maximally described physical state (such as a quantum state) allows for a more detailed description by introducing a mental predicate. In other words, a well-defined physical state turns out to be not a maximal, most detailed description of the system. This is suicide for any physical theory.[21] Thus, it seems necessary to assume the atomicity postulate in a strengthened form: *for any proposition A taken from the wider ordinary language* $\mathscr{L}_0$, *the condition* $A \rightarrow \alpha_b$ *(where* α_b *is an atom in* $\mathscr{L}_b$*) implies either* $\varnothing = A$ *or* $A = \alpha_b$. If this postulate is adopted then the second of the above mentioned alternatives is a forbidden one.

What has been shown above is that the atomicity postulate and the non-equivalence postulate can not be upheld simultaneously. But on the other hand, we have good reason to believe that these two postulates are well-founded. This is what I call fundamental paradox of mind-body relation. In plain language, this means that physical language is 'complete' in the sense that it can describe the system to the maximum precision, yet it cannot be comprehensive enough to tell the 'whole story'. The main purpose of this paper is to show that this paradox can be solved if and only if the usual logic is modified.

For the purpose of deriving the conclusions of the present paper, we do not need to assume the postulate of restricted conjoinability for all member of $\mathscr{L}_m$ and of $\mathscr{L}_b$, as we did in the foregoing. Namely, it suffices to assume that there exist at least one atom α_b of $\mathscr{L}_b$ and at least one member $A_m(\neq \varnothing)$ of $\mathscr{L}_m$ such that $\alpha_b \cap A_m$ is $\varnothing$ or else is neither a physical nor mental proposition and that $\alpha_b{}' \cup A_m$ is $\square$ or else is neither a physical nor mental proposition. This weaker postulate entails also a weaker version of the postulate of non-equivalence. However, the main part of the following will be based on the strong version of the postulate of restricted conjoinability.

4. *Probability Postulates*

It has already been mentioned that we have to assume that we can attach a probability to a conditional proposition of the type: A_m on condition that B_b is true, and B_b on condition that A_m is true. This assumption automatically entails also that we can attach an unconditional probability $p(A_m)$ and $p(B_b)$ to each mental proposition $A_m \in \mathscr{L}_m$ and each physical proposition $B_b \in \mathscr{L}_b$. This is because $p(A_m) = p(A_m|\square)$ and $p(B_b) = = p(B_b|\square)$. Since we have enlarged our language to $\mathscr{L}_0$, it is quite natural to postulate that we can also attach to each proposition A of $\mathscr{L}_0$ an unconditional probability. By probability is usually meant a non-negative quantity such that for any two propositions A and B.

$$p(A) + p(B) = p(A \cap B) + p(A \cup B) \tag{3.1}$$

and

$$p(\varnothing) = 0, \qquad p(\square) = 1. \tag{3.2}$$

We postulate properties (3.1) and (3.2) to be satisfied by what we call unconditional probability. Sometimes, the definition of probability is given by replacing (3.1) by: If $A \cap B = \varnothing$, then $p(A \cup B) = p(A) + p(B)$, and keeping (3.2). It is important to note that (3.1) follows from this latter axiom system only by the use of the distributive law, while the latter axiom system follows from (3.1) without the use of distributive law (See Watanabe [*11*]). It can also be shown that in a finite Boolean lattice an unconditional probability can be attached to each member in such a way that

$$\text{if } A \rightarrow B \text{ and } A \neq B, \qquad \text{then } p(A) < p(B). \tag{3.3}$$

14

To require (3.3) as a general condition for members of our enlarged lattice $\mathscr{L}_0$ may seem somewhat arbitrary but what it represents is a very natural condition. In the first place, whatever philosophical notion of probability is adopted, it is unavoidable to require that if $A \to B$ then $p(A) \leqslant p(B)$. This condition entails automatically that if $A = B$ then $p(A) = p(B)$. The only point implied by (3.3), which may be questioned is the converse of this consequence, namely if $A \to B$ and $p(A) = p(B)$, then $A = B$, or equivalently, if $A \to B$ and $A \neq B$, then $p(A) \neq p(B)$. This is really not too much asked, for it means simply that the probability can be so defined that the difference in A and B shows up in the difference in their probabilities. It seems to me that a philosophical viewpoint is possible that the logical concept is a consequence ensuing from the more basic probabilistic relations. In such a viewpoint, $A = B$ will be interpreted as the condition that it is impossible to give the probability in such a way that $p(A) \neq p(B)$. Conversely, $A \neq B$ will be interpreted as meaning that it is possible to assign probabilities in such a way that $p(A) \neq p(B)$. The issue at the stake in (3.3) is exactly this last condition.

In summary, the following conditions seem to be reasonable to be required. (i) *To each pair (A, B) of propositions in the ordinary language $\mathscr{L}_0$, a conditional probability $p(A \mid B)$ can be attached.* (ii) *To each proposition A in $\mathscr{L}_0$, an unconditional probability $p(A)$ can be attached in such a way that* (3.1) (3.2) *and* (3.3) *are satisfied.* (iii) *If A_m is a mental proposition ($\in \mathscr{L}_m$) and B_b is a physical proposition ($\in \mathscr{L}_b$), then $p(A_m \mid B_b)$ can never be unity except when $A_m = \square$, and $p(B_b \mid A_m)$ can never be unity except when $B_b = \square$.* (The conditional probability with $\varnothing$ as condition is usually indeterminate, and can be dismissed from discussion).

5. Problem of Emergence, Body without Mind, Body with Mind, Mind without Body

By a body without mind is meant an inanimate object which is suspected to have no mind. By a body with mind is meant a certain type of higher animals. It must be clearly understood that the usage of substantives like mind and body does not imply at all that there are 'things' or substances called by these names. This is analogous to the fact that when we speak of wave or flame, we do not mean at all that there are physical matters called wave and flame. They are processes, phenomena or more generally states. Köhler [12] has to be credited for forcefully arguing that most of the existing theories have to rely on a mysterious mechanism of 'emergence' to cope with the fact that there are bodies with mind and bodies without mind. While I do not agree with Köhler's wholesale denunciation of the idea of emergence, I agree with him that this conspicuous difference should not be left unexplained as another logically contingent, empirical fact. Philosophers seem to be often hasty to accept certain things as empirically given and to stop asking further questions. Scientists are more Leibnizian in looking for 'sufficient reasons'. For instance, they want to have some 'explanation' even for the apparently irreducible attributes of an elementary particle, such as the mass of the electron, the spin of the photon. In their so-called explanation, they place themselves in the frame of thought in which the value of a given quantity (such as the mass of an elementary particle) can be other than the empirically given one, and then by the use of a law-like statement, they try to derive the particular empirical value. In the same way, it is desirable that the mind-body duality is considered in a frame of thought in which body can sometimes appear without being accompanied by mind and sometimes being accompanied by mind, and that some law-like characterization is possible to distinguish these two cases.

At this juncture, I may add one speculation which will certainly make me unpopular among philosophers of this century. Since I recognize a high degree of symmetry between mind and body, the existence of a body without mind seems to suggest the possibility of a mind without body. I do not see any reason why the category of soul, ghost, spirit etc., which in our ordinary language is not at all an appendix of a body, should not find some counterpart in our theoretical scheme. Indeed, there is no positive ground against using these concepts. Hence, it is desirable that our theory also includes the possibility of a mind without body. Even if a mind without body is to be excluded, it is better to place ourselves first in the frame of thought in which the possibility of such a being is allowed and then to strike out this possibility by a law-like statement. This would give a 'why' to its 'non-existence'.

Now, we have to examine more carefully what is meant by the small connectives 'with' and 'without' in expressions like 'body with or without mind', 'mind with or without body'. I take the view that the 'without' in those expressions is used in a way very similar to the expression 'This light is colorless (without color)'. In this case, the phrase 'without color' can be

16

interpreted as meaning that no color is specified since all colors are mixed with equal 'weight'. This is further equivalent to saying that a photon emitted by this light source has equal 'probability' of having any wave length (in the visible region). When we say that 'this light has red color (is with red color)' it is equivalent to saying that the probability of a photon having the wave-lengths near the longest end is high. This kind of interpretation of 'without' is inevitable, if we take the view, as we have been doing, that each of the mental and physical parlances covers the universe of discourse, for it is then impossible that a certain object escapes the description in either parlance. This is simply because within $\mathscr{L}_m$, there exists always the negation A'_m to each mental proposition A_m, such that $A_m \cap A'_m = \varnothing$ and $A_m \cup A'_m = \square$. Hence, any object has inherently a possible description in each parlance.

Under these circumstances, the statement that there is a body without mind has to be interpreted as meaning that there is factually a case in which a certain specific physical description A_b in $\mathscr{L}_b$ is actually satisfied, where A_b is such that the conditional probability $p(X_m \mid A_b)$ gives an equal probability to all possible specific mental propositions (atoms) X_m. Similarly, a mind without body corresponds to a mental proposition A_m in $\mathscr{L}_m$ such that the conditional probability $p(X_b \mid A_m)$ gives an equal probability to all possible, specific (atomic) physical propositions X_b. Finally, a body with mind corresponds to a physical proposition A_b in $\mathscr{L}_b$ such that the conditional probability $p(X_m \mid A_b)$ gives a concentrated weight on a certain region of the mental spectrum. I do not claim that this interpretation of inert matters, higher animals and soul after death is the only one, but it is a possible one. Our theory will thus accommodate a place (maybe a little uncomfortable one!) to each of these three kinds of beings, in this particular interpretation of their nature. For the purpose of easy reference in our discussion in Part II, we shall enumerate in the following the postulates made in Part I. A slightly modified version is used in some of the postulates although the ideas are the same. Some mathematical expressions are used without explanations. The reader will find clarification in Part II.

Postulate of Merger

(M1) $\mathscr{L}_0$ (Ordinary Language: a complemented lattice, not necessarily Boolean).

17

$\mathscr{L}_m$(Mental Parlance): a complemented Boolean sublattice of $\mathscr{L}_0$.

$\mathscr{L}_b$(Physical Parlance): a complemented Boolean sublattice of $\mathscr{L}_0$.

(M2) $\varnothing_m = \varnothing_b = \varnothing_0 \equiv \varnothing$, $\square_m = \square_b = \square_0 \equiv \square$

(M3) $\mathscr{L}_m \neq \{\varnothing, \square\}$, $\mathscr{L}_b \neq \{\varnothing, \square\}$, where $\{\varnothing, \square\}$ is a lattice consisting of only two members, $\varnothing$ and $\square$.

Postulate of Non-Equivalence

(NE1) $\mathscr{L}_m \wedge \mathscr{L}_b = \{\varnothing, \square\}$ (a consequence of Postulate of Merger and Postulate of Restricted Conjoinability)

(NE2) $\mathscr{L}_m \vee \mathscr{L}_b \neq \mathscr{L}_0$

The symbols $\vee$ and $\wedge$ are used in the set-theoretical sense here.

Postulate of Restricted Conjoinability

(RC1) $\left. \begin{cases} A_m \in \mathscr{L}_m, & A_m \neq \square \\ B_b \in \mathscr{L}_b, & B_b \neq \square \end{cases} \right\}$ implies

$\left. \begin{cases} \text{a. } A_m \cap B_b \notin \mathscr{L}_b \text{ and } A_m \cap B_b \notin \mathscr{L}_m \text{ or} \\ \text{b. } A_m \cap B_b = \varnothing \end{cases} \right\}$

and dual.

(RC2)   Equivalent to (RC1) and will be explained in Part II.

$\{X \to A_m, \quad A_m \in \mathscr{L}_m, \quad X \in \mathscr{L}_0, \quad X \neq \varnothing, \quad A_m \neq \square\}$ implies $X \notin \mathscr{L}_b$,

$\{X \to B_b, \quad B_b \in \mathscr{L}_b, \quad X \in \mathscr{L}_0, \quad X \neq \varnothing, \quad B_b \neq \square\}$ implies $X \notin \mathscr{L}_m$,

and dual.

Probability Postulate

(P1)   Each pair (A_m, B_m), $A_m \in \mathscr{L}_m$, $B_b \in \mathscr{L}_b$ can be assigned a conditional probability $p(A_m \mid B_b)$ in such a way that for all A_m, $C_m \in \mathscr{L}_m$ and all $B_b \in \mathscr{L}_b$

(a) $p(A_m \mid B_b) + p(C_m \mid B_b) = p(A_m \cap C_m \mid B_b)$
$\quad + p(A_m \cup C_m \mid B_b)$,

(b) $p(\varnothing \mid B_b) = 0$, $p(\square \mid B_b) = 1$,

(c) $p(A_m \mid B_b) \neq 1$ except the case $A_m = \square$ and the case $B_b = \varnothing$.

(P2)   Item (P1) with subscripts m and b interchanged.

18

(P3)   Each $A \in \mathcal{L}_0$ can be assigned an unconditional probability in such a way that for all $A, B \in \mathcal{L}_0$

(a) $p(A) + p(B) = p(A \cap B) + p(A \cup B)$

(b) $p(\square) = 1, \qquad p(\varnothing) = 0$,

(c) If $A \to B$ and $A \neq B$, then $p(A) < p(B)$,

(d) $p(A_m \mid \square) = p(A_m), \qquad p(B_b \mid \square) = p(B_b)$.

Atomicity Postulate

(A1)   There exists at least one $\alpha_b \in \mathcal{L}_b$ such that for any $X_b \in \mathcal{L}_b$ the relation $X_b \to \alpha_b$ implies $\varnothing = X_b$ or $X_b = \alpha_b$.

(A2)   If α_b satisfies the condition (A1), then for any $X \in \mathcal{L}_0$, the relation $X \to \alpha_b$ implies $\varnothing = X$ or $X = \alpha_b$.

Weak Postulate of Restricted Conjoinability

(WRC1) There exist at least one $\alpha_b \in \mathcal{L}_b$ satisfying (A1) and (A2) and at least one $A_m \in \mathcal{L}_m$, $(A_m \neq \varnothing)$, such that (a) $A_m \cap \alpha_b \notin \mathcal{L}_m$ and $\notin \mathcal{L}_b$ or else $= \varnothing$ and (b) $A_m \cup \alpha_b' \in \mathcal{L}_m$ and $\notin \mathcal{L}_b$ or else $= \square$.

Weak Postulate of Non-Equivalence

(WNE1) There exists an $A \in \mathcal{L}_0$, such that $A \in \mathcal{L}_m$ and $A \notin \mathcal{L}_b$. There exists an $A \in \mathcal{L}_0$, such that $A \notin \mathcal{L}_m$ and $A \in \mathcal{L}_b$.

Postulate of Three Modes of Existence (speculation)

(TM1)   Body with mind: There exist pairs (A_m, B_b), $A_m \in \mathcal{L}_m$, $B_b \in \mathcal{L}_b$, $A_m \neq \square$, $B_b \neq \varnothing$, such that $p(A_m \mid B_b)(\neq 1)$ is considerably larger than $p(A_m)$.

(TM2)   Body without mind: There exists $C_b \in \mathcal{L}_b$, such that for all atoms α_m of $\mathcal{L}_m$, $p(\alpha_m \mid C_b)$ is a constant.

(TM3)   Mind without body: There exists $C \in \mathcal{L}_0$ such that for all atoms α_b of $\mathcal{L}_b$, $p(\alpha_b \mid C)$ is a constant.

PART II

MATHEMATICAL ELABORATION AND SOLUTION OF THE PROBLEM

1. *Ordinary Language is Non-Boolean*

In this Part, it will be shown first that our basic postulates laid down in

Part I in conjunction with a very few postulates of minor importance lead to the conclusion that the ordinary language is a non-Boolean, modular lattice. It is not seriously contended that the actual ordinary language is such a lattice, however, it is contended that the conspicuous features extracted from the ordinary language, formulated in a certain formal way, lead to this conclusion. Hence, what follows is a 'model theory', but it shows in later Sections that the fundamental paradox is soluble in this model without being 'dissolved'.

Mathematically speaking, the following argument involves three kinds of numbers: (a) the number of propositions included in a parlance or in a language, (b) the maximum number of 'dimensions' of a proposition in a parlance or a language, (the term dimension will be explained later) and (c) the number of elements that can be combined by simultaneous conjunction or disjunction. Each mathematical theorem used in the following has been proved with a certain condition on these numbers. In order to be mathematically on the safe side, we shall restrict ourselves to the case where the number of propositions in each of the mental and physical parlances is finite (no matter how large) and the number of propositions in the ordinary language is (continuously) infinite, and the number of maximum dimensions in the ordinary language is finite and, the number of elements simultaneously conjoinable is finite. This simplified model already shows that the ordinary language is non-Boolean. This is sufficient to show that a more general model (such as one with infinite dimension numbers) is also non-Boolean for the simplified model can be considered as a special case of the more general model. The case of a Boolean lattice which allows countably many conjunctions is well-known as a σ-algebra. The case of continuously many dimensions is actually used constantly by physicists, but its rigorous treatment has to refer to what von Neuman calls continuous geometry [13]. Some pertinent consideration regarding the relation between the above mentioned three numbers of lattices and the Booleanity of the corresponding logic is given by Birkoff [14]. The method I shall use to prove that the ordinary language is a non-Boolean modular lattice is not found in Birkoff's textbook [14], but most of the basic mathematical theorems that are used in our argument will be found there.

In the following, statements which are necessary or important for the mathematical construction of the model will be numbered by Arabic numerals, while explanations and remarks of philosophical or mathematical nature will be numbered by the lower case Roman numerals. Very often, a mathematical theorem whose rigorous proof can be found elsewhere [14, 15, 16] will be given some illustrations or an informal proof under the Roman numerals. The remarks made in a footnote at the beginning of this paper may be repeated: the metalanguage used in discussing the nature of the ordinary language is Boolean.

(1) Let $\mathscr{L}_m$, called mental parlance, be the set of all mental propositions, which we denote by capital Latin letters with subscript m, such as A_m, B_m ... etc. $\mathscr{L}_m$ is supposed to be closed with respect to the operations of conjunction, disjunction and negation, and the usual laws of a complemented Boolean lattice are assumed to be obeyed. $\mathscr{L}_m$ includes therefore also two elements $\varnothing_m$ and $\square_m$ such that for any $A_m \in \mathscr{L}_m$, the relations $\varnothing_m \to A_m$ and $A_m \to \square_m$ hold, where the relation $\to$ is called implication or entailment and is such that $A_m \to B_m$ is equivalent to $A_m = A_m \cap B_m$.

(i) By a proposition is meant something that can be represented by a symbol and that can be true or false. The expression of a 'proposition' as a sentence is not necessarily adequate. The basic mental proposition is what is phenomenally given to me now. By the virtue of its being phenomenally given to me now, it is considered to be true. A proposition which was true at another time and which, when shifted to the present moment, does not apply, is a false proposition. That which can be given phenomenally to another person is also a mental proposition, hence it is supposed to be true or false. However, the ground of verification of such a mental proposition is not quite clear. Some sort of inference is required, but this inference may be somewhat different from the usual deductive and inductive inference. We, however, do not discuss this problem here. A dispositional, mental proposition is one which cannot be tested directly in the same way as what is or could be 'phenomenally given', but from which such directly testable propositions can be (categorically or probabilistically) derived. A dispositional, mental proposition is also supposed to belong to $\mathscr{L}_m$. The proposition: 'he is very often sleepy' is a dispositional, mental proposition, for it cannot be tested in one instance, but it implies that the probability is high that he is sleepy at an arbitrary instant. We know actually very little about the structure of the relations among mental propositions. However, the fact that we usually assume

21

that it 'makes sense' to use only mental propositions or only physical propositions may be interpreted as meaning that the usual laws of Boolean logic are applicable in each of the two sets of propositions. (1) covers part of Postulate (M1).

(2) Let $\mathscr{L}_b$, called physical parlance, be the set of all physical propositions, which we denote by capital Latin symbols with suffix b, such as A_b, B_b, ..., etc. ('b' for body or behavior). $\mathscr{L}_b$ is supposed to be closed with respect to the operations of conjunction, disjunction and negation which are supposed to obey the usual laws of a complemented, Boolean lattice. $\mathscr{L}_b$ includes also two elements $\varnothing_b$ and $\square_b$ such that for any $A_b \in \mathscr{L}_b$ we have $\varnothing_b \to A_b \to \square_b$. [This last formula means $\varnothing_b \to A_b$ and $A_b \to \square_b$, and not $(\varnothing_b \to A_b) \to \square_b$].

(ii) If we include observational propositions of quantum mechanics in the physical parlance, $\mathscr{L}_b$ becomes necessarily non-Boolean [15, 16]. However, in the first approximation, we may consider $\mathscr{L}_b$ to be Boolean. It will not be so difficult to restore the non-Booleanity of $\mathscr{L}_b$, and the conclusion will remain essentially the same. (2) covers part of Postulate (M1).

(3) For simplicity, it is assumed that the number of elements of $\mathscr{L}_b$ is finite. This entails that there exists a finite number of atoms, α_b, β_b, γ_b, ... etc., none of which is $\varnothing$, and which are disjoint ($\alpha_b \cap \beta_b = \varnothing_b$, etc.) and 'complete' ($\alpha_b \cup \beta_b \cup \gamma_b \cup \ldots = \square$), so that any member of $\mathscr{L}_b$ can be expressed as a disjunction of some atoms, and that for any $X_b \in \mathscr{L}_b$, the relation $X_b \to \alpha_b$ implies $\varnothing_b = X_b$ or $X_b = \alpha_b$. Similarly, it is assumed that the number of elements of $\mathscr{L}_m$ is finite.

(iii) This assumption satisfies the (weak) postulate of atomicity. See # 3, Part I, or Postulate (A1). However, the postulate of atomicity in the form given in Part I does not necessarily imply that $\mathscr{L}_b$ has a finite number of members. For instance, in classical mechanics, a point in the phase space (an Euclidian space defined by spatial coordinates and conjugate momenta) is an atom in the sense that there is no more detailed description, but the number of such points is continuously many. In such a case, two 'types' of languages may be needed to clarify the entire situation, one referring to points in the phase space and the other referring to sets (of non-zero measure) of points. The usual probability theory refers to the latter type of language. Fortunately, in the finite case, these two 'types' of languages become identical. Note: the term 'complete' is used in (3) in a meaning different from in Part I.

(4) It is assumed that $\varnothing_m = \varnothing_b$ and $\square_m = \square_b$.

(iv) This was assumed in Part I as part [Item (i)] of the postulate of non-equivalence, or (M2). We use the symbol $\varnothing$ for both $\varnothing_m$ and $\varnothing_b$ and the symbol $\square$ for both $\square_m$ and $\square_b$.

(5) $X_m \epsilon \mathscr{L}_m$ implies (a) $X_m = \varnothing$, or (b) $X_m = \square$, or (c) $X_m \notin \mathscr{L}_b$. Similarly, $X_b \epsilon \mathscr{L}_b$ implies (a) $X_b = \varnothing$, or (b) $X_b = \square$, or (c) $X_b \notin \mathscr{L}_b$.

(v) This is a mathematical expression of [Item (ii) of] the postulate of non-equivalence in Section 2 of Chapter I and is equivalent to (NE1).

(6) The ordinary language $\mathscr{L}_0$ is a set of propositions such that there exists a transitive, one-way relation (implication) $A \to B$ for some pairs (A, B) of its member-propositions and it contains two elements $\varnothing$ and $\square$ which satisfy the relation $\varnothing \to X \to \square$ whenever $X \epsilon \mathscr{L}_0$. ($\mathscr{L}_0$ is a partially ordered set).[22]

(vi) This is not a complete definition of $\mathscr{L}_0$, but only one of the conditions of $\mathscr{L}_0$. The conjunction $C = A \cap B$ of two elements A and B of $\mathscr{L}_0$ is defined as an element of $\mathscr{L}_0$ satisfying the following two conditions: (a) $C \to A$, $C \to B$ and (b) the relations $X \to A$, $X \to B$, $X \epsilon \mathscr{L}_0$ imply $X \to C$. The disjunction $C = A \cup B$ of two elements A and B of $\mathscr{L}_0$ is defined by (a) and (b) in which the arrows $\to$ are replaced by arrows $\leftarrow$. With an additional assumption of the unique existence of $C \epsilon \mathscr{L}_0$, (6) establishes that $\mathscr{L}_0$ is a lattice, [part of Postulate (M1)], which implies that the operations of conjunction and disjunction in $\mathscr{L}_0$ satisfy the idempotent law $[A \cap A = A, A \cup A = A]$, the commutative law $[A \cap B = B \cap A, \quad A \cup B = B \cup A]$, the associative law $\left[A \cap (B \cap C) = (A \cap B) \cap C, A \cup (B \cup C) = (A \cup B) \cup C\right]$ and the absorptive law $\left[A \cap (A \cup B) = A, A \cup (A \cap B) = A\right.$. It is of vital importance to note that the distributive law $\left[A \cap (B \cup C) = (A \cap B) \cup (A \cap C), \quad A \cup (B \cap C) = (A \cup B) \cap (A \cup C)\right]$ does not follow from the definition. The assumption (6) represents the first part of the postulate of merger. It can be easily shown by the use of the definition of $\cap$ and $\cup$ that $A \to B$ is equivalent to $A = A \cap B$ and to $B = A \cup B$.[22]

(7) The ordinary language $\mathscr{L}_0$ is such that if A is a member of $\mathscr{L}_0$, then its negation of A' is also a member of $\mathscr{L}_0$, where the relation between A and A' is such that the law of double negation, the law of contraposition, the law of self-contradiction hold.

(vii) This establishes that $\mathscr{L}_0$ is a complemented lattice. Part of (M1). The law of double negation means that $(A')' = A$. The law of contraposition

23

states that if $A \to B$ then $B' \to A'$. The law of self-contradiction states that if $A \to A'$ then $A = \varnothing$. The reason why we cannot define the operation of negation by two laws: $A \cap A' = \varnothing$ and $A \cup A' = \square$ is explained in Watanabe's text book [16]. Conversely, if A' is the negation of A, then these two laws hold. From (7) follows de Morgan's law: $(A \cap B)' = A' \cup B'$. From (6) and (7) follows that the relation $A \to B$ implies, but is not necessarily implied by, $A' \cup B = \square$.

(8) The ordinary language $\mathscr{L}_0$ is such that $\mathscr{L}_m$ and $\mathscr{L}_b$ are its sublattices so that the conjunction, disjunction and negation taken within each of $\mathscr{L}_m$ and $\mathscr{L}_b$ remain the conjunction, disjunction and negation defined in $\mathscr{L}_0$.

(viii) This is the main part of the postulate of merger, see Postulate (M1).

(9) Let $A_m \in \mathscr{L}_m$, $A_b \in \mathscr{L}_b$, and $X \in \mathscr{L}_0$. The ordinary language $\mathscr{L}_0$ is such that the relations $A_m \to X$, $A_m \neq \varnothing$, $X \neq \square$ imply $X \notin \mathscr{L}_b$ and that $A_b \to X$, $A_b \neq \varnothing$, $X \neq \square$ imply $X \notin \mathscr{L}_m$.

(ix) From (9) follows also that the relations $X \to A_m$, $X \neq \varnothing$, $A_m \neq \square$ imply $X \notin \mathscr{L}_b$. The premise implies, in virtue of the law of contraposition that $A'_m \to X'$ and $A'_m \in \mathscr{L}_m$, $X' \in \mathscr{L}_0$, $A'_m \neq \varnothing$, $X' \neq \square$. Hence, by the use of (9) we conclude that $X' \notin \mathscr{L}_b$. This last condition implies $X \notin \mathscr{L}_b$. For, if $X \in \mathscr{L}_b$ then $X' \in \mathscr{L}_b$. Similarly, the relation $X \to A_b$, $X \neq \varnothing$, $A_m \neq \square$ imply $X \notin \mathscr{L}_m$. Conversely, the two consequences derived here from (9) imply (9). Hence, the two sets of conditions are equivalent. (9) is (RC2).

(10) The condition (9) is equivalent to the condition that $A_m \cap B_b$ with $A_m \neq \square$ and $B_b \neq \square$ [$A_m \cup B_b$ with $A_m \neq \varnothing$, $B_m \neq \varnothing$] does not belong to either $\mathscr{L}_m$ or $\mathscr{L}_b$ except when it is equivalent to $\varnothing$ [$\square$].

(x) The condition introduced in (10) is the main part of the postulate of restricted conjoinability. First, let us assume (9) and prove the new condition introduced in (10). Since $\mathscr{L}_0$ is closed for conjunction, $C = A_m \cap B_b$ is a member of $\mathscr{L}_0$. By the definition of conjunction, $C \to A_m$ and $C \to B_b$. It suffices to prove that C is bound to be $\varnothing$ or else to belong to neither $\mathscr{L}_m$ nor $\mathscr{L}_b$. Suppose $C \neq \varnothing$, then from the relation $C \to A_m$ and the condition $A_m \neq \square$ follows, according to (ix), that $C \notin \mathscr{L}_b$. Similarly, it follows that $C \notin \mathscr{L}_m$. Next, let us assume the new condition of (10) and derive the condition (9). The premise $A_m \to X$, is equivalent to $A_m = A_m \cap X$, where $X \neq \square$, hence $A_m \neq \square$. According to (10), if $X \in \mathscr{L}_b$, then $A_m \cap X$ cannot belong to either $\mathscr{L}_m$ or $\mathscr{L}_b$, except in the case $A_m \cap X = \emptyset$. But this is impossible, in view of $A_m = A_m \cap X$, because by the premise of (9) we have $A_m \neq \emptyset$. Hence $X \notin \mathscr{L}_b$. (10) establishes equivalence of (RC1) and (RC2).

(11) The condition (5) is a consequence of condition (9) or equivalently, of the conditions mentioned in (10).

(xi) The proof was already given in Part I.

(12) The set theoretical union $\mathscr{L}_R = \mathscr{L}_m \vee \mathscr{L}_b$ satisfying the conditions so far mentioned about the ordinary language is called Rylean language.

(xii) The adjective Rylean is used as a reminder that Ryle did not like to admit the propositions which do not belong to either $\mathscr{L}_m$ or $\mathscr{L}_b$, and this does not necessarily imply that $\mathscr{L}_R$ is the kind of language Ryle is discussing in his book.

(13) Theorem: The Rylean language which contains at lease one proposition which is neither $\emptyset$ nor $\square$ in each parlance $\mathscr{L}_m$ and $\mathscr{L}_b$ is non-Boolean.

(xiii) *Proof:* Let A_m and B_b be such that $A_m \in \mathscr{L}_m$, $A_m \neq \emptyset$, $A_m \neq \square$, and $B_b \in \mathscr{L}_b$, $B_b \neq \emptyset$, $B_b \neq \square$. Then, we can conclude that $A_m \cap B_b = \emptyset$ for any A_m and B_b satisfying these conditions. The reason is that according to (10), $A_m \cap B_b$ cannot belong to $\mathscr{L}_m$ or $\mathscr{L}_b$, except when it is $\emptyset$, but there is no element in $\mathscr{L}_R$ except those which belong to $\mathscr{L}_m$ or $\mathscr{L}_b$. Similarly, $A_m \cup B_b = \square$ for any A_m and B_b satisfying the above conditions. Since $A_m \in \mathscr{L}_m$, we have by definition also $A'_m \in \mathscr{L}_m$, and $A'_m \neq \emptyset$, $A'_m \neq \square$. Consider $A'_m \cup (A_m \cap B_b)$ and $(A'_m \cup A_m) \cap (A'_m \cup B_b)$. If these two are not equivalent, then the distributive law is violated, i.e., $\mathscr{L}_R$ is non-Boolean. In fact, $A'_m \cup (A_m \cap B_b) = A'_m \cup \emptyset = A'_m$ (the last step is justified in any lattice since it is equivalent to $\emptyset \to A'_m$) and $(A'_m \cup A_m) \cap (A'_m \cup B_b) = \square \cap \square = \square$ (idempotent law). Since $A'_m \neq \square$, we conclude that $\mathscr{L}_R$ is non-Boolean. It may be noted that we have not used the postulate of atomicity here.

(14) The ordinary language $\mathscr{L}_0$ is non-Rylean, i.e., it contains at least one element which is not in $\mathscr{L}_R = \mathscr{L}_m \vee \mathscr{L}_b$.

(xiv) The Rylean language is an outrageously poor language, hence it is natural to admit some elements which are neither mental nor physical. This was already required in the postulate of restricted conjoinability in Section 3 of Part I and was repeated as (NE2). It is, however, not specific-

25

ally decided here whether $\mathscr{L}_0$ consists only of those elements that can be formed from elements of $\mathscr{L}_m$ and $\mathscr{L}_b$ by a repeated use of disjunction, conjunction, negation. About this matter, we postulate only (8).

(15) The ordinary language $\mathscr{L}_0$ is such that if α_b is an atom of the physical parlance, then there exists in $\mathscr{L}_0$ no element X such that $X \rightarrow \alpha_b$ does not imply $\varnothing = X$ or $X = \alpha_b$. (Similarly for an atom α_m of the mental parlance).

(xv) In other words, for any $X \in \mathscr{L}_0$ the relation $X \rightarrow \alpha$ implies $\varnothing = X$ or $X = \alpha$, where α is any mental or physical atom. This is what we called [strong form of] postulate of atomicity in Section 3 of Part I and was repeated as Postulate (A2). Actually, existence of atoms in $\mathscr{L}_m$ and satisfaction of (15) by them are not necessary for our following argument.

(16) Theorem: The ordinary language is non-Boolean, provided that each of $\mathscr{L}_m$ and $\mathscr{L}_b$ contains at least one element which is neither $\varnothing$ nor $\square$.

(xvi) Proof: Take an atom α_b in the physical parlance $\mathscr{L}_b$ and take any mental proposition A_m which is neither $\varnothing$ nor $\square$. Consider $C = A_m \cap \alpha_b$. This C is a member of $\mathscr{L}_0$ and satisfies $C \rightarrow \alpha_b$, hence $C = \varnothing$ or $C = \alpha_b$ The latter possibility is excluded, since it would imply $\alpha_b = A_m \cap \alpha_b$, which would violate the condition (10), since α_b cannot be $\square$. Now consider $\alpha'_b \cup (A_m \cap \alpha_b)$ and $(\alpha'_b \cup A_m) \cap (\alpha'_b \cup \alpha_b)$. The first term is equivalent to $\alpha'_b \cup \varnothing = \alpha'_b$. The second term is equivalent to $(\alpha'_b \cup A_m) \cap \square = \alpha'_b \cup A_m$. Now, if $\alpha'_b = \alpha'_b \cup A_m$ then the distributive law (for this particular case) is satisfied. But, since $\alpha'_b \neq \varnothing$, $A_m \neq \varnothing$, we obtain from (10) that $\alpha'_b \cup A_m \notin \mathscr{L}_b$ or else $\alpha'_b \cup A_m = \square$. The both eventualities contradict the relation $\alpha'_b = \alpha'_b \cup A_m$, which implies $\alpha'_b \cup A_m \in \mathscr{L}_b$ and $\alpha'_b \cup A_m \neq \square$. (This latter is due to the condition $\alpha_b \neq \varnothing$). Hence $\alpha'_b \neq \alpha'_b \cup A_m$, which implies that the distributive law does not hold. Q.E.D. The proof becomes slightly simpler if we rewrite $\alpha'_b = \alpha'_b \cup A_m$ as $A_m \rightarrow \alpha'_b$ and use (9) instead of (10). It should be noted that atomicity condition (15) is used only for $\mathscr{L}_b$. To a frame of mind which is accustomed only to the usual logic, a non-Boolean structure appears as a paradox. That is why we stated that our postulates lead to a paradox, but they do not involve any internal contradiction in a broader framework of logic.

(17) Theorem: Let $\mathscr{L}_1$ and $\mathscr{L}_2$ be two sub-lattices of a lattice $\mathscr{L}_0$. If there exist an $A_1 \in \mathscr{L}_1$ and a $B_2 \in \mathscr{L}_2$ satisfying the following three

26

conditions, then $\mathscr{L}_0$ is a non-Boolean lattice. (a) $A_1 \neq \varnothing$, $B_2 \neq \varnothing$; (b) $A_1 \cap B_2 = \varnothing$; (c) $A_1 \cup B'_2 = \square$ or $A_1 \cup B'_2 \notin \mathscr{L}_2$.

(xvii) We can prove this theorem exactly in the same way as in (xvi) just by replacing A_m and α_b by A_1 and B_2 respectively. The reader can easily see that (13) and (16) are corollaries to this theorem. Provided that the atomicity condition (15) for $\mathscr{L}_b$ is assumed, we can therefore still conclude the non-Booleanity of ordinary language $\mathscr{L}_0$ by replacing (9) and (10) by a much weaker postulate: there exists an atom α_b of $\mathscr{L}_b$ and a member $A_m(\neq \varnothing)$ of $\mathscr{L}_m$ such that $\alpha_b \cap A_m$ is $\varnothing$ or else does not belong to $\mathscr{L}_b$ and that $\alpha'_b \cup A_m$ is $\square$ or else belongs to neither $\mathscr{L}_b$ nor $\mathscr{L}_m$. This is the reason why (WRC1) is sufficient to conclude the non-Booleanity of $\mathscr{L}_0$. We have seen that the postulate of non-equivalence (5) follows from the postulate of restricted conjoinability (9). If (9) is replaced by the above-stated weaker postulate, the postulate (5) does not follow. What follows from this weaker postulate is that there exists an $A_m \in \mathscr{L}_m$ which satisfies $A_m \notin \mathscr{L}_b$ and that there exists a $B_b \in \mathscr{L}_b$ which satisfies $B_b \notin \mathscr{L}_m$.

The class of non-Boolean lattices is exessively broad, hence, it is desirable to locate the ordinary language in a smaller class, which is non-Boolean, yet not too liberal. The conditions we postulated about probabilities will serve the purpose of reasonably restricting the class.

2. Ordinary Language is Modular

(18) The ordinary language $\mathscr{L}_0$ is such that we can attach to each member A a real number $p(A)$ in such a way that (a) $p(A) + p(B) = p(A \cap B) + p(A \cup B)$, and (b) $p(\square) = 1$, $p(\varnothing) = 0$ and (c) if $A \to B$ and not $B \to A$, then $p(A) < p(B)$.

(xviii) This is part of Item (ii) of our probability postulates introduced in Section 4 of Part I and corresponds to (a), (b), (c) of (P3).

(19) From the condition that $\mathscr{L}_0$ is a lattice (not necessarily Boolean) follows that if $A \in \mathscr{L}_0$, $B \in \mathscr{L}_0$, $C \in \mathscr{L}_0$ and $A \to C$, then $A \cup (B \cap C) \to (A \cup B) \cap C$.

(xix) Proof: Call $X \equiv A \cup (B \cap C)$ and $Y \equiv (A \cup B) \cap C$. What is to be proven is $X \cup Y = Y$. The following transformations involve only those laws which are obeyed by any general lattice. First $X \cup Y = A \cup [(B \cap C) \cup Y]$ by associative law. Now call $Z = B \cap C$, then $Z \cap Y = Z$, because $Z \cap Y = B \cap C \cap (A \cup B) \cap C = [B \cap (A \cup B)] \cap C = B \cap C = Z$. This means $Z \to Y$, i.e., $Z \cup Y = Y$. Hence, $X \cup Y = A \cup (Z \cup Y) =$

$= A \cup Y$. By the use of the premise $A = A \cap C$, this becomes $X \cup Y =$
$= (A \cap C) \cup [(A \cup B) \cap C]$. But, $(A \cap C) \cap [(A \cup B) \cap C] =$
$[A \cap (A \cup B)] \cap C = A \cap C$ meaning $A \cap C \to (A \cup B) \cap C$ which
is Y. Hence $X \cup Y = (A \cap C) \cup Y = Y$. Q.E.D.

(20) Property (a) of (18) ensures that $p[A \cup (B \cap C)] = p[(A \cup B)$
$\cap C]$ provided $A \to C$.

(xx) *Proof:* The left hand side can be transformed as $p[A \cup (B \cap C)] =$
$p(A) + p(B \cap C) - p(A \cap B \cap C) = p(A) + p(B) + p(C) - p(B \cup C)$
$- p(A \cap B \cap C)$, where the last term can be written as $p(A \cap B)$ since
$A \cap C = A$. But $p(A \cap B) = p(A) + p(B) - p(A \cup B)$. Hence,
$p[A \cup (B \cap C)] = p(C) + p(A \cup B) - p(B \cup C)$. Now, the right hand
side is transformed as $p[(A \cup B) \cap C] = p(A \cup B) + p(C) -$
$- p(A \cup B \cup C)$, where the last term is, on account of $A \to C$, equal to
$p(B \cup C)$, thus $p[(A \cup B) \cap C] = p(A \cup B) + p(C) - p(B \cup C)$.
Q.E.D.

(21) *Dedekind's Theorem:* A lattice satisfying the condition (18) (hence
the ordinary language) is a modular lattice, where a modular lattice is a
lattice satisfying the condition that if $A \in \mathscr{L}_t$, $B \in \mathscr{L}_0$, $C \in \mathscr{L}_0$ and $A \to C$,
then

$$A \cup (B \cap C) = (A \cup B) \cap C. \qquad (2.1)$$

(xxi) *Proof:* From (19), we have the result that if $A \to C$ then $X \to Y$,
where X and Y are the left and right sides of (2.1). From (20) we have the
result that any p satisfying condition (a) of (18) satisfies the relation
$p(X) = p(Y)$ provided $A \to C$. Since $X \to Y$, the case $X \ne Y$ is excluded,
for $X \ne Y$ implies then, in virtue of condition (b) of (18), that $p(X) < p(Y)$.
Hence $X = Y$. Q.E.D.

(22) The modular law is admission of the Boolean (distributive) law
$(A \cap C) \cup (B \cap C) = (A \cup B) \cap C$ for the case $A \to C$.
(xxii) Put $A = A \cap C$ in (2.1).

(23) The modular law is self-dual.
(xxiii) The two distributive laws pass from one to the other by the dual
transformation, but the modular law comes back to itself by the dual
transformation.

(24) In a modular lattice, the distributive laws hold if one of the three
elements involved in them implies one of the remaining two elements.
(xxiv) The proof must be relegated to my text book [16].

(25) In a modular lattice, there exists always a Boolean sublattice.

(xxv) The set $\{\varnothing, \square\}$ is one. $\{\varnothing, A, A', \square\}$ is another. There can be many more.

(26) A Boolean lattice is a modular lattice.

(xxvi) If the distributive laws hold for any three elements, then the modular law holds for any three elements satisfying the condition $A \to C$.

(27) The ordinary language $\mathscr{L}_0$ is a non-Boolean, modular lattice.

(xxvii) The last Section has proved that $\mathscr{L}_0$ is non-Boolean and the present Section has proved that $\mathscr{L}_0$ is modular. However, in order to prove the non-Booleanity we have used several postulates. It remains to be proven that these postulates can be satisfied within the framework of a modular lattice. This will be done in the next Section.

3. *Geometrical Model and Consistency of Postulates*

In this section, we introduce a geometrical model for the non-Boolean, modular lattice in order to show that all the postulates introduced in Part I can be satisfied by the model, hence they are compatible with the claim that the ordinary language is a non-Boolean, modular lattice.

(28) Consider an n-dimensional Euclidean space, and let $\mathscr{L}_0$ be the set of all subspaces passing the origin. $\mathscr{L}_0$ includes the origin itself $\varnothing$ and the entire space $\square$.

(xxviii) If $n = 3$, any straight line passing the origin (one-dimensional subspace) and any plane passing the origin (two-dimensional subspace) are members of $\mathscr{L}_0$. The zero-dimensional subspace (the origin), $\varnothing$, and the 3-dimensional subspace (the entire space), $\square$, are also members of $\mathscr{L}_0$. Each proposition in the ordinary language corresponds to one of the subspaces in the n-dimensional space and vice versa. The number of members in $\mathscr{L}_0$ is obviously continuously infinite.

(29) Each subspace A is a set of vectors $\vec{\Psi}$ attached to the origin such that if $\vec{\Psi}_1$ and $\vec{\Psi}_2$ belong to A then any linear combination $a_1\vec{\Psi}_1 + b\vec{\Psi}_2$ (a_1, a_2, arbitrary real numbers) belongs again to A.

(xxix) This is a property of subspaces which the reader can easily check for the case $n = 3$.

(30) A subspace A is said to 'imply' a subspace B, $(A \to B)$, if and only if the following condition is satisfied: if a vector $\vec{\Psi}$ belongs to A then it belongs to B. Also, $\varnothing \to A \to \square$, for any A. (Hence $\mathscr{L}_0$ is a partially ordered set).

(xxx) Thus, for instance, if A is a straight line lying on the plane B, then

29

$A \rightarrow B$. The relation $\varnothing \rightarrow A$ may be interpreted as meaning that the origin lies on any subspace. The relation $A \rightarrow \square$ is obvious.

(31) By the use of the definition introduced in (vi), the conjunction of two subspaces A and B becomes the set of vectors which belong to both A and B. (This set is a subspace). Similarly, the disjunction of two subspaces A and B becomes the set of all linear combinations of two vectors, one from A and the other from B. ($\mathscr{L}_0$ is a lattice).

(xxxi) The conjunction is easy to understand. For instance, if A and B are two planes, then $A \cap B$ is the intersection of these two planes which is a straight line. The disjunction of two straight lines A and B becomes a plane which contains both of these straight lines A and B. The reason is as follows. The subspace (straight line) A implies A itself, and any plane on which A lies, and also $\square$. Similarly, straight line B implies itself, and any plane on which B lies and also $\square$. The common elements which are implied by both A and B are then the plane on which both A and B lie and $\square$. The smallest of them is of course the plane passing both A and B.

(32) The dimension of a subspace A is the maximum number of vectors that can be taken in A such that they are linearly independent. We denote by $D(A)$ the dimension of A.

(xxxii) m vectors $\vec{\Psi}_1, \vec{\Psi}_2, \ldots, \vec{\Psi}_m$ are said to be linearly independent if the equation $a_1\vec{\Psi}_1 + a_2\vec{\Psi}_2 + \ldots + a_m\vec{\Psi}_m = 0$ can be satisfied only by $a_1 = a_2 = \ldots = a_m = 0$. But the reader may be satisfied with his knowledge that the dimension of $\varnothing$ is zero, the dimension of a straight line is one, the dimension of a plane is two, and the dimension of $\square$ (in the 3-dimensional space) is three.

(33) If A is m-dimensional, B is l-dimensional and $A \cap B$ is k-dimensional then $A \cup B$ is $(m + l - k)$-dimensional.

(xxxiii) A rigorous proof can be found in reference 16. The reader may be satisfied by noticing two cases. Two different planes A and B in a three dimensional space have a straight line $A \cap B$ as their intersection. $D(A) = 2$, $D(B) = 2$, $D(A \cap B) = 1$. The disjunction of A and B is according to (31) is the entire space. Hence $D(A \cup B) = 3 = 2 + 2 - 1$, agreeing with (33). If two planes coincide, then, of course, $A \cap B$ and $A \cup B$ are also the same plane. Hence, $D(A) = D(B) = D(A \cap B) = D(A \cup B) = 2$ which again satisfies the claim of (33).

(34) If $A \rightarrow B$ and $A \neq B$, then $D(A) < D(B)$.

(xxxiv) If subspace A is m-dimensional then subspace B implied by A must be either m-dimensional or more-than-m-dimensional. But if B is m-dimensional then B must coincide with A. Hence B must be more-than-m-dimensional.

(35) The lattice $\mathscr{L}_0$ of subspaces in modular.

(xxxv) This is a consequence of Dedekind's theorem (21) using the D-function of (33) and (34). An illustration will presently be given that modular law (2.1) is in fact satisfied.

(36) Take three distinct straight lines A, B, C in the same plane P, then $A \cup (B \cap C) = A \cup \varnothing = A$ and $(A \cup B) \cap (A \cup C) = P \cap P = P$. Hence, the lattice is non-Boolean.

(xxxvi) To prove the non-Booleanity of lattice $\mathscr{L}_0$, it suffices to adduce one example of breakdown of the distributive law. That is what is done in (36). Take another example. Let C be the x-y-plane and B be the z-axis. Let A be an arbitrary straight line, which neither coincides with the z-axis nor lies in the x-y-plane. Then, $A \cup (B \cap C) = A$, because $B \cap C = \varnothing$. On the other hand, $(A \cup B) \cap (A \cup C) = D \cap \square = D$, where D is the plane defined by the z-axis and the straight line A. Hence, the distributive law breaks down again. Now suppose, in this example, that the straight line A lies on C, $(A \rightarrow C)$. Then, $A \cup (B \cap C)$ is still A, but $(A \cup B) \cap (A \cup C)$ becomes $D \cap C$ which becomes the A again. Thus, if $A \rightarrow C$, then the distributive law $A \cup (B \cap C) = (A \cup B) \cap (A \cup C)$ holds since $A \cup C = C$. This is the modular law. The reader can check the distributive law holds in the illustration of (36) if the straight lines A and C coincide ($A = C$ hence $A \rightarrow C$).

(37) An unconditional probability of a subspace A can be defined by $D(A)/n$.

(xxxvii) Then, we have satisfied all the requirements of (3.1) (3.2) and (3.3) of Part I. [Items (ii) of Probability Postulate of Section 4 or (a), (b), (c) of (P3) of Section 5].

(38) The negation A' of subspace is the set of all vectors each of which is perpendicular to all the vectors in A. If $A = \varnothing$, then $A' = \square$.

(xxxviii) In the case of $n = 3$, if A is a straight line A' is the plane whose normal is A. If A is a plane, then A' is its normal. Adding to this the definition $\varnothing' = \square$ and $\square' = \varnothing$, the reader can check that all three basic laws, viz., the law of double negation, the law of contraposition and the law of self-contradiction hold.

31

(39) $A \cap B = \varnothing$ and $A \cup B = \square$ do not guarantee that $B = A'$. (xxxix) Take as A the x-y-plane and take as B a straight line which neither coincides with the z-axis nor lies in A.

(40) Take a system of n rectangular coordinates in the space, and consider the set $\mathscr{B}$ of all subspaces which are defined by some of the coordinates. $\mathscr{B}$ is called a family of orthogonal subspaces. There are continuously many $\mathscr{B}$'s in the space.

(xl) In the case $n = 3$, for each chosen rectangular coordinate system, $\mathscr{B}$ consists of eight elements, $\varnothing$, x-axis, y-axis, z-axis, x-y-plane, y-z-plane, z-x-plane, and $\square$. Each coordinate system corresponds one-to-one to a $\mathscr{B}$, and there are continuously many ways of taking a coordinate system.

(41) Each $\mathscr{B}$ is a Boolean sublattice of $\mathscr{L}_0$.

(xli) First $\mathscr{B}$ is obviously a complemented lattice, using the same definition of conjunction, disjunction and negation as in $\mathscr{L}_0$, and each member of $\mathscr{B}$ is a member of $\mathscr{L}_0$. The rest is to prove that $\mathscr{B}$ is Boolean. The reason why $\mathscr{B}$ is Boolean is that $\mathscr{B}$ is isomorphic to the class of subsets taken from the set of n objects which corresponds to the n coordinates. In plain words, each member A of $\mathscr{B}$ corresponds one-to-one to a subset of the set of n coordinates, in such a way that $A \cap B (A \cup B)$ corresponds to the set theoretical conjunction (disjunction) of two subsets of coordinates corresponding respectively to A and B. But, the class of subsets of a set is obviously Boolean. The reader can confirm in each of the examples of violation of the distributive law mentioned above, a small modification which makes all subspaces involved belong to a single $\mathscr{B}$ and results in compliance with the distributive law. This includes the example in (xxxix).

(42) There are two Boolean sublattices $\mathscr{L}_m$ and $\mathscr{L}_b$ in $\mathscr{L}_0$ such that they have only $\varnothing$ and $\square$ in common.

(xlii) Each of $\mathscr{L}_m$ and $\mathscr{L}_b$ corresponds to a $\mathscr{B}$ hence to a coordinate system. The condition mentioned in (42) can be easily realized by taking the two coordinate systems $\mathscr{L}_m$ and $\mathscr{L}_b$ is such a way that no proper subspace (i.e. subspace which is neither the origin nor the entire space) belonging to one coordinate system coincides with a proper subspace of the other coordinate system. In the three-dimensional space, we need only take the two coordinate systems in such a way that no coordinate axis of one system coincides with any coordinate axis of the other system and no coordinate plane of one system coincides with any coordinate plane of the other system. (42) satisfies (M1), (M2) and (NE1).

(43) There are two Boolean sublattices $\mathscr{L}_m$ and $\mathscr{L}_b$ in $\mathscr{L}_0$ such that if $A_m \in \mathscr{L}_m$, $A_m \neq \square$, and $B_b \in \mathscr{L}_b$, $B_b \neq \square$, then $A_m \cap B_b \notin \mathscr{L}_m$ and $A_m \cap B_b \notin \mathscr{L}_b$ or else $A_m \cap B_b = \varnothing$.

(xliii) This corresponds to (RC1). This condition can be satisfied by adding a further condition to those already mentioned in (xlii), namely, that no proper subspace of one system should be a subspace of a proper subspace of the other system. The conditions mentioned in (xlii) and this additional condition can be satisfied by requiring that each coordinate of one system should have a non-vanishing component along each of the coordinate axes of the other system. In the three-dimensional space, we should require other than the conditions in (xlii) that no coordinate axis of one system should lie in a coordinate plane of the other system. Suppose that we take the x'-axis of the x'-y'-z'-system outside any coordinate plane of the x-y-z-system. If we take the y'-axis and the z'-axis on the plane perpendicular to the x'-axis in such a way that neither of them lies in any of the coordinate planes of the x-y-z-system, then both (42) and (43) are satisfied. If we take the y'-axis in the x-y-plane, then (42) is still satisfied but (43) is non longer satisfied. This shows that the postulate of non-equivalence (NE1) is a weaker requirement than the postulate of restricted conjoinability (RC1). Consider further the case where the x-y-plane and the x'-y'-plane coincide but the x'-axis does not coincide with either x-axis or y-axis. In this case both the postulate of restricted conjoinability, (43) and (RC1), and the postulate of non-equivalence, (42) and (NE1), are violated. But, on the other hand, the weak postulate of restricted conjoinability (WRC1) and the weak postulate of non-equivalence (WNE1) are still satisfied. Even in this weakest case, a lattice of which $\mathscr{L}_m$ and $\mathscr{L}_b$ are sublattices cannot be Boolean. Of course, (WRC1) and (WNE1) are satisfied in the previous two cases. See below, (xliv), for the definition of atoms.

(44) Each atom α of $\mathscr{L}_m$ as well as of $\mathscr{L}_b$ is such that $\varnothing \to X \to \alpha$, $X \in \mathscr{L}_0$ imply $\varnothing = X$ or $X = \alpha$.

(xliv) The atoms of a $\mathscr{B}$ is obviously the one-dimensional subspaces (straight lines) which belong to $\mathscr{B}$. For $n = 3$, the x-axis, y-axis, and z-axis are the atoms of the $\mathscr{B}$ corresponding to the x-y-z-coordinate system. Now there is no subspace in the entire space which is smaller than (i.e.,

33

which implies and is not equal to) a one dimensional subspace except $\varnothing$. Hence $\varnothing \to X \to \alpha$ implies $\varnothing = X$ or $X = \alpha$ if α is a one-dimensional subspace. This satisfies the stronger form of the atomicity postulate. (A2)

(45) Hence the postulate of restricted conjoinability and the postulate of atomicity can both be satisfied. The paradox is resolved.

(xlv) Thus, we have proven that the postulate of merger, the postulate of non-equivalence, the postulate of atomicity and the postulate of unconditional probability are all upheld without interfering with one another. All that remains to be shown is that the postulate of conditional probabilities can be satisfied and that the problem of body without mind and mind without body can be given some kind of explanation.

4. *Conditional Probabilities*

We shall show in this Section that we can define the conditional probability $p(A_m \mid B_b)$ of $A_m \in \mathscr{L}_m$ on condition that $B_b \in \mathscr{L}_b$ is true, in such a way that $p(A_m \mid B_b)$ never becomes unity except when $A_m = \square$ or $B_b = \varnothing$ and $p(B_b \mid A_m)$ never becomes unity except when $B_b = \square$ or $A_m = \varnothing$, and that the unconditional probabilities $p(A_m)$ and $p(B_b)$ defined in $\# 3$ become special cases $p(A_m \mid \square)$ and $p(B_b \mid \square)$, satifying the remaining part of Probability Postulate.

(46) Let $\vec{\varphi}$ be the projection of a vector $\vec{\Psi}$ on the subspace A, and write $\vec{\varphi} = \mathscr{P}[A]\vec{\Psi}$, where $\mathscr{P}[A]$ is called projection operator of A. The matrix expression of $\mathscr{P}[A]$ has the following properties which do not depend on the coordinate system used to express $\mathscr{P}[A|$ as the matrix. (a) It is symmetric. (b) It is idempotent $\mathscr{P}[A]\,\mathscr{P}[A] = \mathscr{P}[A]$. (c) Each diagonal element is non-negative and not larger than unity. (d) The sum of all diagonal elements is the dimension of A, Spur $\mathscr{P}[A] = D(A)$.

(xlvi) For the case $n = 3$, let us first take as A a straight line whose direction cosines are a_1, a_2, a_3. Let $\vec{\Psi}$ be a vector whose components are Ψ_1, Ψ_2, Ψ_3, and let the components of $\vec{\varphi}$ be φ_1, φ_2, φ_3. It is obvious that $\vec{\varphi}$ has the same direction as the straight line A. The length of $\vec{\varphi}$ is $a_1\Psi_1 + a_2\Psi_2 + a_3\Psi_3$, hence its x-component, φ_1 is $\varphi_1 = a_1(a_1\Psi_1 + a_2\Psi_2 + a_3\Psi_3)$. In general, $\varphi_i = \sum_{j=1}^{3} a_i a_j \Psi_j$, $(i = 1, 2, 3)$. This means that the matrix corresponding to the projection operator is $(\mathscr{P}[A])_{ij} = a_i a_j$. All properties (a) (b) (c) (d) are clearly satisfied. Property (b) has a simple geometrical meaning: if we project $\vec{\varphi}$ once more on A, we get again $\vec{\varphi}$. When A is a plain, we take two mutually orthogonal directions on it, and take the sum of the two projection matrices corresponding to these two directions. By a simple calculation, one can convince oneself that $\mathscr{P}[A]$ thus obtained has the property of projecting any vector $\vec{\Psi}$ onto its projection $\vec{\varphi}$ on the plane A, in the sense that $\varphi_i = \sum_{j=1}^{3} (\mathscr{P}[A])_{ij} \Psi_j$.

Properties (a) (b) (c) remain unchanged by addition of two matrices corresponding to the two orthogonal directions lying in A. As far as the diagonal sum is concerned, it becomes simply the sum of the diagonal sums of the two matrices. Hence Spur $(\mathscr{P}[A]) = 2 = D(A)$. It can also be shown that $\mathscr{P}[A]$ actually does not depend on the choice of the two orthogonal directions arbitrarily taken on A.

(47) If we take two arbitrary subspaces A and B, then in general $\mathscr{P}[A]\mathscr{P}[B]$ and $\mathscr{P}[B]\mathscr{P}[A]$ are not equal to each other. However, if A and B belong to the same family of orthogonal subspaces then $\mathscr{P}[A]\mathscr{P}[B] = \mathscr{P}[B]\mathscr{P}[A]$.

(xlvii) Take two planes A and B which do not intersect each other at the right angle. Then, the result of projection of a vector $\vec{\Psi}$ on A first and then on B is different from the result of projecting $\vec{\Psi}$ on B first and then on A. If A and B belong to the same $\mathscr{B}$, then they intersect each other at the right angle and the result of both operations is equivalent to projecting $\vec{\Psi}$ on the straight line which is the intersection of A and B.

(48) The projection on $\varnothing$ is multiplication by 0, and the projection on $\square$ is multiplication by 1.

(xlviii) Any vector projected on the origin becomes the origin itself. Any vector projected on the entire space becomes itself.

(49) If A and B are orthogonal, i.e., if each vector of A is orthogonal to all vectors of B, then $\mathscr{P}[A]\mathscr{P}[B] = \mathscr{P}[B]\mathscr{P}[A] = 0$.

(xlix) If we project a vector on B it becomes a vector in B which is orthogonal to A, i.e., its further projection on A becomes zero.

(50) If A' is the negation of A, then $\mathscr{P}[A]\mathscr{P}[A'] = \mathscr{P}[A']\mathscr{P}[A] = 0$ and $\mathscr{P}[A] + \mathscr{P}[A'] = 1$.

(l) If we project a vector $\vec{\Psi}$ on a plane and on its normal, and if we add these two resulting vectors we obtain $\vec{\Psi}$ back.

(51) The conditional probability of A on B can be defined by $p(A \mid B) =$ Spur $(\mathscr{P}[A]\mathscr{P}[B])/D(B)$.

(li) The matrix product $\mathscr{P}[A]\mathscr{P}[B]$ does not necessarily have the properties of a projection operator, but the diagonal sum is definable.

(52) The quantity $p(A \mid B)$ defined in (51) has the following propositions. (a) $p(A \mid B) \geqslant 0$, (b) $p(\varnothing \mid B) = 0$, (c) $p(\square \mid B) = 1$, (d) $p(A \mid B) + p(A' \mid B) = 1$.

(lii) Property (b) follows from (48). Property (c) follows from (48) and Spur $\mathscr{P}[B] = D(B)$. Property (d) follows from (50) and Spur $\mathscr{P}[B] = D(B)$. Property (a) for two straight lines A and B is obvious from the matrix expressions of $\mathscr{P}[A]$ and $\mathscr{P}[B]$. Actually, it is easy to see that $p(A \mid B)$ for two straight lines A and B becomes $\cos^2 \theta$, where θ is the angle between the two lines. If A is a straight line and B is a plane, then $p(A \mid B) = \cos^2 \theta/2$, where θ is the angle between A and B. If A is a plane and B is a straight line then $p(A \mid B) = \cos^2 \theta$. If A and B are both planes, then $p(A \mid B) = (1 + \cos^2 \theta)/2$ where θ is the angle between the two planes.

(53) The conditional probability defined in (51) does not satisfy the condition that $p(A \mid C) + p(B \mid C)$ should be equal to $p(A \cap B \mid C) + p(A \cup B \mid C)$ unless A and B belong to the same $\mathscr{B}$.

(liii) Take as A the x-y-plane, and take as B the plane passing the x-axis making angle θ with the x-y-plane. Let C be the z-axis. Then $p(A \mid C) = 0$, $p(B \mid C) = \sin^2 \theta, p(A \cap B \mid C) = 0, p(A \cup B \mid C) = 1$. Hence the relation $p(A \mid C) + p(B \mid C) = p(A \cap B \mid C) + p(A \cup B \mid C)$ is not satisfied unless $\theta = 90^0$. This relation holds also for $\theta = 0$, because then $p(A \cup B \mid C) = 0$. This shows that the conditional probabilities can be used as the usual probabilities so long as the main events (the proposition X of $p(X / Y)$) are limited to the same $\mathscr{B}$.

(54) If $B = \square$, then $p(A \mid B) = p(A)$, where $p(A)$ is the quantity defined in (37).

(liv) If we put $\mathscr{P}[B] = 1$ in the definition of (51), then $p(A \mid B) =$ Spur $\mathscr{P}[A]/D(B) = D(A)/n$. It is interesting that the unconditional probability satisfies the condition $p(A) + p(B) = p(A \cap B) + p(A \cup B)$ even if A and B do not belong to the same $\mathscr{B}$. Thus, (a) (b) of (P1) and (d) of (P3) are satisfied.

(55) If $B \to A$ then A and B belong to the same $\mathscr{B}$, and $p(A \mid B) = 1$, provided $D(B) \neq 0$.

36

(lv) If $B \rightarrow A$ and if B is m-dimensional and A is k-dimensional, then we can take the k orthogonal directions defining A in such a way that m out of the k defines B. Hence, A and B belong to the same family of orthogonal spaces. And, it is obvious that any vector projected on B first and then on A, or on A first and then B, becomes simply its projection on B. That is, if $B \rightarrow A$ then $\mathscr{P}[B]\mathscr{P}[A] = \mathscr{P}[A]\mathscr{P}[B] = \mathscr{P}[B]$. Hence from the definition of (51), $p(A \mid B) = 1$, provided Spur $\mathscr{P}[B] = D(B) \neq 0$.

(56) If $p(A \mid B) = 1$, then $B \rightarrow A$.

(lvi) First we note that if B does not imply A then $\mathscr{P}[A]\mathscr{P}[B] \neq \mathscr{P}[B]$. We shall show that if $\mathscr{P}[A]\mathscr{P}[B] \neq \mathscr{P}[B]$ then Spur $(\mathscr{P}[A]\mathscr{P}[B]) \neq$ Spur $\mathscr{P}[B]$. Since the diagonal sum does not depend on the coordinate system, we can choose the latter in such a way that $\mathscr{P}[B]$ becomes diagonal and has 0's and 1's on the diagonal. Then, the condition $\mathscr{P}[A]\mathscr{P}[B] \neq \mathscr{P}[B]$ means that $\mathscr{P}[A]$ has a value different from 1 at least at one place where $\mathscr{P}[B]$ has 1. Since a diagonal element of $\mathscr{P}[A]$ can never be larger than 1, those values which are different from 1 are less than 1. Hence, Spur $(\mathscr{P}[A]\mathscr{P}[B]) <$ Spur $\mathscr{P}[B]$. This means that $p(A \mid B) < 1$. The reader can check for the case of $n = 3$ by using the expressions (in terms of cosines) of $p(A \mid B)$ we introduced in (lii).

(57) We can always take two families $\mathscr{B}$'s, say $\mathscr{L}_m$ and $\mathscr{L}_b$, such that $p(A_m \mid B_b) \neq 1$ except for $A_m = \square$ or $B_b = \varnothing$ and $p(B_b \mid A_m) \neq 1$ except for $B_b = \square$ or $A_m = \varnothing$.

(lvii) All we need is to take the coordinates of $\mathscr{L}_m$ and coordinates of $\mathscr{L}_b$ as we did in (xliii) to satisfy (43). Then, no member B_b of $\mathscr{L}_b$ implies a member A_m of $\mathscr{L}_m$ except in the case $A_m = \square$. Hence, by the use of (56), we conclude that $p(A_m \mid B_b) \neq 1$. Similarly, for $p(B_b \mid A_m) \neq 1$. Item (c) of (P1) is satisfied. It is easy to see that item (iv) of the postulate of Section 2, Part I can also be satisfied. It may be noted that if we adopt (WRC1) instead of (RC1), then (c) of (P1) is not necessarily true. This is because some subspace of one coordinate system can become a subspace of a proper subspace of the other coordinate system.

5. *Body with Mind, Body without Mind, Mind without Body*

In spite of the fact that we denied the possibility of implication of the type $A_m \rightarrow B_b (B_b \neq \square)$ and $B_b \rightarrow A_m (A_m \neq \square)$, we have to admit that there are pairs of mental and physical propositions A_m, B_b such that there is a very high probability $p(B_b \mid A_m)$ of B_b being true when A_m is known to

37

be true, or vice versa. Such a situation is certainly taking place in most of animals with mind. See (TM1). This is the main reason why the equivalence hypothesis enjoyed a strong support among philosophers. But on the other hand, in a case like an inanimate object, we have to assume that the object corresponds to a physical state B_b such that it gives evenly distributed probabilities to all mental propositions. With the help of the concept of atoms, this condition can be formulated as meaning that there is a physical state B_b such that the conditional probability $p(\alpha_m \mid B_b)$ is equal for all atoms α_m of $\mathscr{L}_m$. See (TM2). In the same way, a mind without body must correspond to a mental proposition A_m such that the conditional probability $p(\alpha_b \mid A_m)$ is equal for all atoms α_b of $\mathscr{L}_b$. The question is now; is it possible for $\mathscr{L}_m$ and $\mathscr{L}_b$ to contain all these three cases.

In terms of the geometrical model, we may ask the question: is it possible to choose two coordinate systems $\mathscr{L}_m$ and $\mathscr{L}_b$ such that some coordinates of $\mathscr{L}_b$ have components of equal length along all the coordinates of $\mathscr{L}_m$ and some other coordinates 'almost' imply some subspaces of $\mathscr{L}_m$. Let (x, y, z) be the rectangular coordinates of $\mathscr{L}_m$ and let (x', y', z') be the rectangular coordinates of $\mathscr{L}_b$. Suppose we take the x'-axis in the direction whose direction cosines are $(1/\sqrt{3}, 1/\sqrt{3}, 1/\sqrt{3})$, and the y'-axis in the direction $(-1/\sqrt{6}, -\sqrt{1/6}, \sqrt{\frac{2}{3}})$ and the z'-axis in the direction $(-1/\sqrt{2}, 1/\sqrt{2}, 0)$. Then, the physical proposition B_b corresponding to the x'-axis gives the conditional probability $p(\alpha_m \mid B_b) = \frac{1}{3}$ to all three atomic mental propositions, satisfying (TM2), while the physical proposition C_b corresponding to the z'-axis strictly imply the mental proposition D_m corresponding to the x-y-plane. Of course this choice of the x'-y'-z'-coordinates violates the postulate of restricted conjoinability, (RC1) .[23] But, now we can rotate the x'-y'-z'-coordinates very slightly about the z'-axis, then $p(\alpha_m \mid B_b)$ still remains $\frac{1}{3}$ for all α_m and $p(D_m \mid C_b)$ becomes slightly less than unity, satisfying (TM1). This is the proof in our model of possible existence of a 'body without mind' and a 'body with mind' with a strong mind-body correlation.

It is true in the simple model, which has a room for a body without mind, but does not have a room for mind without body. However, even in this model there exist vectors in $\mathscr{L}_0$ which make an equal angle to all physical axes, satisfying (TM3). We can reserve such a vector for a 'soul' without mind, although such a state of pure soul is not quite in the mental language. Actually considerable flexibility is gained (a) by increasing the number of dimensions and (b) by taking into account that $\mathscr{L}_b$ itself is already a non-Boolean language, hence probably $\mathscr{L}_m$ too. In such a broad framework, it will become much easier to accommodate all kinds of delicate mind-body relations. A question which is entirely unanswered in the above simple geometrical model is: why does a particular pair of coordinate systems bear the important meaning of mental and physical parlances, while there are infinitely many pairs which satisfy the same mutual relations.

6. *Concluding Remarks*

I do not mean to attach any realistic meaning to the geometrical model of the non-Boolean lattice which I used in the last sections. It was used in order to show that all the required postulates are compatible with the basic assumption that the ordinary language is essentially a non-Boolean, modular lattice.

The main goal achieved by this assumption is a synthesis, which is otherwise impossible, of the thesis that the most detailed physical description of a person cannot leave anything more to describe and the thesis that no pair of mental and physical propositions can be quite equivalent to each other, not only with respect to their connotation but also with respect to their denotation.[24] But, this synthesis was so made that the possibility of approximate equivalence between a mental proposition and certain physical propositions is upheld. Exact equivalence between a mental proposition and a certain physical proposition has been assumed by many philosophers but has never been convincingly proven. Approximate equivalence, in the form adopted here, can accommodate various recondite ideas nurtured by 'tender-minded' philosophers, without getting into logical inconsistency.

IBM Research Center, Yorktown Heights, New York

NOTES

1. The facts that give rise to the mind-body problem are essentially of empirical nature. Acceptance of certain propositions of empirical nature, however, leads us to abandon the narrow framework of logic we usually impose on ourselves. I take the view that logic has to be modified or improved if it is necessary in order to coordinate empirical facts in a comprehensible form. When we introduce a new type of logic, the metalanguage used in explaining the new logic is supposed to

comply with the old logic. This is permissible because the new logic is such that the old logic retains its validity in a limited universe of discourse.

2. In spite of the intended neutrality toward metaphysical doctrines, the model theory developed here seems to lead naturally to the view that mind as well as body is nothing more than a 'function' or 'state' and that the 'thing' which is supposed to be performing the 'function' or is supposed to be in the 'state' possesses absolutely no attributes (nothingness!), for the description of the function or state exhausts all there is to be described.

3. It may be of interest to note that to an (imaginary) intelligence which is incapable of apperceiving a geometrical situation beyond a two-dimensional world, this conjunction will appear to be either inconceivable or sheer non-sense.

4. A little more rigorous formulation of the paradox is found in Section 3 of this Part I, and the most rigorous formulation is found under heading (16) in Section I of Part II.

5. This sense of the term model is familiar to natural scientists, for any theory in a natural science is a model in this sense. There is not a single real gas which satisfies Boyle-Charles' law or Van der Waals' law, yet these laws are useful models. So are the scores of 'models' of atomic nuclei.

6. This sense of the term model is familiar to logicians. A 'model' is used by them in order to demonstrate the absence of internal contradictions in an abstract system defined by a set of postulates. The above-mentioned requirements correspond in this analogy to the 'postulates'.

7. For an interesting and relevant discussion of the notion of 'goal' in machines, see Scheffler [18].

8. At the N.Y.U. meeting I associated my idea of dual alternatives to Bohr's idea of complementarity but, this association was found to be misleading, since my idea is considerably more precise and elaborate than the idea of complementarity, albeit that the former is strongly influenced by the latter.

9. In some version of the postulate 'a certain' should be replaced by 'each'.

10. The mental propositions include those which stand for what is phenomenally given to me, and those which are supposed to represent what is phenomenally given to other people as well as those which express a dispositional, mental proposition like 'he is often sleepy'.

11. The relation between Feigl's theory and the present one is such that the latter shares the ontological simplicity of the former in spite of the fact that the latter does not assume the two conceptual systems to be extensionally identical as the former does without sufficient proof.

12. The thesis of the equivalence postulate is of course a law-like proposition, however, its instantiation whose experimental verification should 'confirm' it becomes a statement of this type: $A_m \rightleftarrows A_b$.

13. We are interested in feasibility in principle and not in a feasibility conditioned by irrelevant factors.

14. Let (XY) stand for a proposition that Y is true *and* immediately thereafter X is true. Then, the equality of the probability of $(A_m A_b)$ to that of $(A_b A_m)$ for all cases establishes the absence of the mutual interference discussed here. See S. Watanabe [16].

15. Read $p(A_m \mid A_b)$ as the probability of A_m on the condition that A_b is true. Similarly for $p(A_b \mid A_m)$. The relation $p(A_m \mid A_b) = 1$ is equivalent to the statement that A_b implies A_m except in the case where the probability of A_b is zero.

16. The reader can confirm for himself when he comes to the later Sections of this paper that our proposal of a new logic fulfils these two requirements perfectly.
17. Any process of explication involves a danger of losing sight of some important aspects of the subject-matter for it tends to select what one wants to see in it.
18. Such as the claim that the mind-body problem involves a paradox of logical nature and that it can be resolved by introducing a non-Boolean logic.
19. The usage of the term 'atom' in this paper conforms with that of mathematicians, but not with that of some philosophers.
20. It is of course not meant that the actually used ordinary language is such a thing. This is a 'model' in the 'first' sense of the word.
21. This would lead to the so-called hypothesis of 'hidden parameter' which was repudiated by Von Neumann [17].
22. The relation $A \to B$ implies always the relation $A \supset B = A' \cup B = \square$, but the latter implies the former only if the distributive law holds. Thus, implication ($\to$) and material implication ($\supset$) must be distinguished. The usual propositional calculus based on $\supset$, which has a built-in distributive law, has to be considered as having a limited validity.
23. This however satisfies the weak postulate of restricted conjoinability (WRC1).
24. In the weaker version of our postulates, (WRC1) and (WNE1), some, *but not all*, mental propositions can have physical counterparts which are exactly equivalent to them. However, even in this case, the non-Booleanity of logic has to be introduced to achieve the synthesis in question.

REFERENCES

[1] Ryle, G., *The Concept of Mind*, London, 1949.
[2] Watanabe, S., *Annals of Japan Assoc. for Philosophy of Science* 1, 4, 12.
[3] Watanabe, S., *Louis de Broglie, physicien et penseur* (éd. André George), Paris, Albin Michel, 1952, p. 385.
[4] Wigner, E. P., A lecture entitled *Two Kinds of Reality*, presented at Marquette University, June 1961 (to be published).
[5] Watanabe, S., *Dimensions of Mind*, (ed. Sidney Hook), New York University Press, New York, 1960, p. 143.
[6] Feigl, H., *Dimensions of Mind*, p. 24.
[7] Putnam, H., *Dimensions of Mind*, p. 148.
[8] Bohr, N., *Atomic Physics and Human Knowledge*, John Wiley, New York, 1958
[9] Hampshire, S. *Mind* 14 (1950) 237.
[10] Ducasse, C., *Dimensions of Mind*, p. 85.
[11] Watanabe, S., *Inference and Information*, John Wiley, New York, (in preparation), Chapter VI.
[12] Köhler, W., *Dimensions of Mind*, p. 3.
[13] von Neumann, J., *Continuous Geometry*, Princeton University Press, Princeton, 1960.
[14] Birkoff, G., *Lattice Theory*, (rev. ed.) Amer. Math. Soc., New York, 1948.
[15] Birkoff, G., and von Neumann, J., *Ann. Math.* 37 (1936) 823.
[16] Watanabe, S., *Inference and Information*, Chapter V.
[17] von Neumann, J. *Mathematische Grundlagen der Quantenmechanik*, Julius Springer, Berlin, 1932.
[18] Scheffler, I., *Brit. Journ. Philos. Sci.* 9 (1959) 265.

COMMENTS

RAY SOLOMONOFF

The comments that I will make will be almost entirely on the first part of Dr. Watanabe's paper, which gives some basis for the postulates that he makes. My own view on this is that the mind-body problem is an unnecessary problem. I think that it really can be avoided, and that there are some advantages in so doing.

The first postulate that was used was derived from the following difficulty: We start out with two languages, a "body language" and a "mind language". Either the mind language has some essentially new material in it that is not in the body language (or the physical language), or it does not. If it does have some new material, this contradicts one of our basic ideas of quantum mechanics – that the state-function does indeed contain all of the information. If it does not, then the physical language can express everything that the mind language can express – and perhaps even more.

Dr. Watanabe feels that the second possibility is unsatisfactory for two reasons. One is that it is operationally impossible to make the correspondence between the two languages, viz. between a specific mental state and a purported quantum mechanical state that we want to make it correspond to. The other reason is that we feel that in some sense the mental language is not a redundant one – that it does have something essential to offer; and that if the physical language is complete, this would contradict our intuitive ideas about this matter.

First of all, I'll try to show that we can make the correspondence between the physical language and the mind language without any difficulties due to quantum mechanics.

Then I will try to show that while this makes the mind language redundant in a formal sense, it still leaves the mind language as useful and as necessary to the progress of science as ever before.

Formally stated, the correspondence problem is as follows: Suppose we have a normal human being as a subject and he has certain mental states that he himself can identify. If we have a theory of a correspondence between mental and physical states, then this theory can be represented by a large table that lists a set of physical states of the person in the first column, and gives names of the corresponding mental states of the person in the second column. We want to find out whether this table is correct.

One thing that we might do is to first observe the physical state of this person. Having observed it we look up his physical state in the table, and we tell him what mental state it corresponds to. We then ask him: "Well, were you in that state at that time?" He will say yes or no, and this will verify or negate our theory.

It is Dr. Watanabe's contention that this in general will not be possible, because we cannot observe the physical state of a person exactly. In fact, some of the significant processes in the brain happen on the quantum level, so that a few quanta used by the observer can, in some cases, produce an unknown, though significant change in the state of the organism.

Although it is not essential to the argument I will use, I will mention as an aside, that there is some reason to believe that the "macro-operation" of the human brain may not be significantly disturbed by a few quanta of energy. This ability to operate properly in a background of disturbances (if indeed this ability exists) may be accounted for by particular kinds of error correcting circuits in the brain.

The idea that a few quanta may produce important changes in the brain, stems, perhaps, from human response to very low levels of light – on the order of a few quanta. Such sensitivities, however, occur very rarely, and only after long periods of acclimatization to darkness. The problem of making a correspondence between mental and physical states is not significantly modified if we consider a man who is temporarily cut off from external stimuli.

However, let us return to the main argument – suppose the quantum mechanical disturbances in observation *are* important. How can we go about verifying our table of correspondences? First we give the person some sort of visual input – a bright light which we focussed on his retina. Then we compute through quantum mechanics what his physical state should be (at this point, we note that this is just a *Gedankenexperiment*). We can't tell *exactly* what state he is in: say we have them narrowed down to ten possible states, and we get a theoretical probability distribution over these states. We then go to the table and get a probability distribution for the corresponding mental states. We then give the list of possible mental states to the person who's being experimented on: we ask him which one of these he's experienced, and he tells us.

We do this experiment many times with many different kinds of inputs. Eventually we will be able – not to verify whether a particular mental state corresponds to a particular physical one – but we will be able to verify the *table as a whole*, which is quite another thing.

This corresponds, to some degree, to the fact that while we cannot tell what state a particular sodium atom is in at any particular time, we can verify various facts about sodium atoms in general to a high degree of accuracy. While we can accurately verify our general theory of the structure of the sodium atom, we cannot be certain as to how things worked out in the case of any particular atom.

In a similar way, we may not be able to verify in any particular case the direct correspondence between certain specific physical and mental states – but we *can* verify *as a single theory*, the entire table of correspondences.

The next point I will discuss is the apparent resultant redundancy of the "mind language". Suppose we do find that our physical language is adequate and that we can make this correspondence and it appears that our mind language is, at best, redundant. Due to this redundancy, shouldn't we throw out this mind language? The answer is that we can if we want to, but I don't think we will or should. This is because it is of great value to us. First of all, heuristically, it enables us to work in a very direct manner with many concepts that would not be suggested by the purely quantum mechanical picture. We are used to working with this particular language, and can make quick and easy inductions with it. It will suggest correspondences between states that we could never conceive of otherwise.

Perhaps an analogy would clarify this point. At the present time the science of chemistry as we ordinarily know it is "redundant". All of its information content can be more compactly expressed as a small set of quantum mechanical equations. All of the literature of chemistry can be viewed as a development, *ad nauseam*, of this rather simple set of equations.

We are certain, however, that classical non-quantum mechanical chemistry is extremely useful. The development of any practical results from the quantum mechanical equations is at best an arduous process, and is in most cases well beyond the power of our present-day mathematics. The language of classical chemistry is usually very convenient and compact for describing chemical reactions. Using this language it is possible to make good approximate models of chemical reactions. While these approximate models are not as accurate as the quantum mechanical ones, they are far more easily computable, and they are heuristically useful, in the sense that they are readily grasped by the mind of man. This last is of much importance. A model that a man can easily work with, will readily suggest new interesting experiments to him. A more accurate model, that is not so readily mentally manipulated, will tend to be far less suggestive to him.

Another kind of utility for the mind language is that any additional language is useful in induction. Many inductions that would be extremely unlikely using the physical language alone, become quite reasonable if we have the mind language to work with. Empirically, we find that suitably controlled inductions, using the mind language, are as reliable as any other kind of acceptable induction.

My suggestion at this point is that any additional languages for describing the world should not be accepted or rejected on the basis of their redundancy. It is far better to ask of a language: 1) Is it useful in induction? 2) Does it often suggest new experiments or observations that turn out to be interesting? 3) Does it enable us to describe phenomena easily? 4) Does it suggest simple models of the phenomena it describes?

Most scientists find it useful to use several ostensibly different languages in working with phenomena in a given field.

My conclusions are about the same as Dr. Watanabe's, but my reasons for them are somewhat different. I feel that the dualism between the mind language and the physical language is important and desirable and that both have much to contribute in the phenomena of interest. I also feel that the correspondence between mental and physical states can never be certain – but this is because of the practical limitations of finite sample size, rather than any inherent theoretical difficulty.

ZATOR Corporation, Cambridge, Massachusetts

COMMENTS

MARCUS GOODALL

The question of the relation between a physical and a mental language suggests as an analogy the complementarity of identity and interaction. By a very gross analogy you can imagine the investigation of identity is a physical type of investigation, and the investigation of interaction is a mental one. Now this is a very gross analogy, but it's not too bad.

Now the next point – the reason I think it's only approximate (and this is obvious) – is that one can offer something which might be better. If you think of this dualism in another way, it's a dualism of two modes of thought being pushed to the limit – the limit being when you get into problems of self-reference. Now Dr. Watanabe mentioned the problems of self-reference in physics: namely, that when you're discussing an elementary particle you have to do it in terms of something still more elementary. A similar situation is present in mental phenomena, which I prefer to call 'cognitive systems' because that gets away from any anthropomorphism. A cognitive system is something you can imagine has the properties of being intelligent. Now you cannot define, deterministically at any rate, an intelligent machine, because if you do so you get into a contradiction. For if it can be defined, then it has a strategy. But the essential characteristic of intelligence is that it learns to improve its strategy, so it must have a strategy which works on the strategy and so on. This is a typical problem of self-reference. So one has two extreme theories where you get involved in the problems of self-reference.

Now in the case of quantum mechanics, it's known that a modular lattice is not good enough to describe quantum mechanics. It's good enough to describe one-particle theory but not several interacting particles that are created and annihilated, or even, really, in the one-particle case where you have emission and absorption of radiation. This was known to Von Neumann and Birkoff. But what *is* the appropriate logic for quantum mechanics, is not yet known. There are various guesses. One possibility is semimodular, and another possibility is a skew lattice where there is a non-commuting combination of elements.

Is there a possibility of a more general approach? I claim that there is. The central idea is to start from the point of view of how does one deal with the problems of self-reference. This is a problem which came up in logic at the beginning of the century, and Russell and Whitehead offered the theory of logical types to handle it. The theory of logical types has gone through a number of developments; none of them is really quite satisfactory. But what they all imply is an appeal to a many-valued logic. You can see that if you've got a paradoxical state – one which can be neither true nor false – there must be some other value than true or false, and if you look at the attempts to offer some substitute for the theory of types you get things which are weakenings of two-valued logic. They are attempts to open up the system to the generation of types; that is, the introduction of rules about rules which will make the system more solvable. Now you can see what the limitation is in the modular form of logic. It's known that if you try and make a representation of complementary operators or variables in quantum mechanics, the range of values cannot both have the same cardinal number. If one is countable the other must be uncountable, and this means that they are different logical types. So it's a theory with just two logical types. And it seems from the example of second quantum theory that this isn't good enough. For already the state occupation operators have an uncountable range while that of the complementary phase operators is clearly not countable.

Massachusetts Institute of Technology, Cambridge, Massachusetts

ERIC H. LENNEBERG

THE RELATIONSHIP OF LANGUAGE TO THE
FORMATION OF CONCEPTS

SUMMARY OF ORAL PRESENTATION *

Presented November 30, 1961

LANGUAGE AND THE PHYSICAL WORLD

Consider W. v. Orman Quine's distinction between meaning and reference. Determination of meaning in terms of physical description is not feasible. The question remains whether the occurrence of those words that have reference can be fully accounted for by some objective description of physical phenomena. A reductionist semantic theory would answer this affirmatively. It would hold that the lexicon of a natural language has a set of primitive terms whose reference is exhaustively describable in terms of physical specifications (i.e. sensory terms such as *red, hard, square*), and that the rest of the lexicon is in turn describable in terms of these "primitive words". Formal demonstration of the unacceptability of reductionist semantic theories exists but we shall confine our criticisms here to the following few simple observations:

In an empirical test to map the reference of experiential words into spaces defined by strictly physical dimensions it appears that only a few families of words (all color words, two or three words for texture, two or three words for taste, etc.) can be so mapped whereas most other words with seemingly simple references require spaces that are impossible to dimensionalize in any objective physical way. (Think e.g. of *big, dry, far,* etc.). Nor is the second assumption of a reductionist theory empirically true: there is no way of predicting or defining the use of the vast majority of so-called concrete words in terms of sensory words. Neither is it

---

* The oral presentation differed from this summary in that some of the evidence referred to here was given in greater detail. Reference and discussion of my own earlier work is left in rather skeletal form here since the results have been published elsewhere.

possible simply to catalogue exhaustively all the concrete objects that might be covered by such a word as "table" and there is no *objective* set of *physical* criteria that might generally define all conditions ever to be labled "table".

Thus a lexicon does not bear a direct one-to-one relationship to the physical word. Labeling the world by means of words is essentially dependent upon a peculiarly human conceptualization of reality. Since it appears to be impossible to specify the semantics of words by reference to reality, I am proposing as an alternative to specify it in terms of human concepts.

CONCEPT FORMATION AS STUDIED IN THE LABORATORY

If words are said to correspond more exactly to human concepts than to discrete physical phenomena, it behooves us to ask, how are the concepts acquired? We may wish to turn to laboratory work for an elucidation. In the laboratory, concept formation is most commonly studied by requiring school children or adults to discover the criterion for grouping objects. [4] Most interesting in these tests is the fact that some concept formation tasks are considerably easier to accomplish than others. [7, 8] What causes this gradient of difficulty is essentially an unsolved problem even though many explanations have been offered. There is, however, one very salient facilitating factor: if the criterion for grouping stimuli is something like "hats" or "red things" or "five things" subjects attain the concept easily. If the criterion is "all configurations that have at least three but no more than five rounded shapes enclosed in the upper left hand corner of the figures and that have a prime number of dots distributed randomly along the bottom line are examples of the concept", subjects' task is difficult. (In oral presentation it was demonstrated that the difference between the two types of criteria is not simply due to detailed *vs* undetailed criteria). The more difficult it becomes to express the criterion for grouping in the natural language of the subject, the harder is the task. In other words, in most instances of experimental concept formation, there is a correlation between ease of *naming* the concept and ease of attaining it.

Since subjects in these experiments are always in full command of the English language, laboratory results do not necessarily explain either how *words* are learned or concepts formed in the *developing infant.* [12] Thus, this type of experiment by and large begs the question we are interested in: if words match concepts of reality rather than the physical world itself, what is the origin of the concepts? The laboratory experiments on mature subjects heightens our curiosity about this problem but they do not answer it. If concepts are difficult to attain if one does not have a word for them, and if words name concepts, i.e. are learned if there is a concept already present, how can anything ever be learned? The only possible answer to this is an essentially Neo-Kantian one. Man as well as lower animals is so constituted as to order impinging physical stimuli in a pre-determined and species-specific way, leading to a peculiar and characteristic formation of concepts. There is circumstantial evidence for this as follows.

THE QUESTION OF CONCEPT-LANGUAGE PRIORITY

(1) In all cultures infants learn to speak (use words) at the same age, i.e. about 18 to 20 months. Words are not used in exactly the same fashion as in the adult language, showing that there is no "blind imitation" but an underlying principle. But words are never used in a random fashion. To call early words "concrete" and late words "abstract" is a simple convention, not an explanatory principle. Words such as "up", "again", "he" or the use of plurals or the definite article are the quintessence of abstractness from the chimpanzee's viewpoint. Objective specification of the physical reality that is properly referred to by these words would be so difficult that no high speed computer would have enough storage capacity for the details necessary to distinguish acceptable from unacceptable physical instances of these lexical and syntactic "meanings". Yet all infants learn these and thousands of similar ones with ease and with great speed – even children with defective IQ's. [13] Since there is neither time for nor knowledge of how to train "organisms" to use these words, but children seem to accomplish this following a maturational schedule (provided they are exposed to an appropriate environment) it is likely that they do so because they can match the word they are given to a concept that they now have.

(2) If words corresponded to (named) an objectively definable array of physical objects, then the learning of language would be heavily dependent on an intact sensorium in the learner. Vision should be of

greatest importance since it is our main avenue of sensory contact with the physical world. Congenitally blind children should, consequently, have insuperable difficulty in early stages of language acquisition. Empirically this is not true. Congenitally blind children not further handicapped by psychiatric or other neurological disease learn to speak according to the same maturational schedule as sighted children. Even congenital deaf-blindness is no insurmountable obstacle to language attainment. Sensory communication with the physical world must be unquestionably less important to language acquisition than an autochthonous growth of concepts.

(3) One might argue against the last point above that perhaps the acquisition of language precipitates or precedes concept formation in the blind and the deaf-blind. There is evidence against this. The sighted but congenitally deaf child grows up to school age (5 to 6) literally without language. Examination of these children makes it clear that before formal training in special schools on lip-reading, reading and writing, and speech has begun these children have a vocabulary of less than ten words (my own research in progress). Yet observation of their play and behavior reveals ordered, well-planned behavior including complex inferences and combination of thought that makes their activities indistinguishable from hearing children of the same age. Even in non-verbal, formal tests on concept formation and cognitive processes, scores only differ minimally (with a few exceptions, see cited literature), from the normal population.[6,15]

(4) We conclude this list with a refutation of the notion that a certain type of research might be contrary evidence to our hypothesis. Since von Senden's publication of the operated congenital cataract cases, it has been held that cognition, down to sensory perception, is a "learned" phenomenon and that therefore the social environment, particularly language, may help "shape" perception and cognition. However, (1) Neither von Senden's compilation nor the subsequent empirical works are unequivocal.[17] (2) If the development of pattern perception can eventually be shown to be dependent upon stimulation contingencies, the stimulation is certain to be of a purely physical nature. Nowhere do the physical characteristics of man's environment differ sufficiently to allow us postulation of a differential effect upon development of vision. Yet natural languages give the appearance of different semantic structure.[18]

COGNITIVE SIGNIFICANCE OF LANGUAGE DIVERSITY

Since there is no perfect translatability between any two languages, there must be some culture-bound arbitrariness in "naming". To what extent does this arbitrariness affect conceptualization?

(1) Partly, to the extent that languages actually differ from one another. However, in many aspects languages are universally alike even though there is arbitrariness within limits.[11]

(2) The difference between languages can be determined objectively, but only in very special cases.[1, 5, 9, 14]

(3) Influence of particular language features upon certain of their speakers' cognitive aspects *can* be demonstrated. Generally speaking, however, it occurs only under conditions of drastic "cue reduction".[3, 10, 12] In behavior not artificially altered by cue reduction (and not always even in these special instances) language and concepts do not seem to be significantly interdependent.

TOWARDS A THEORETICAL FORMULATION

Both language and human cognitive processes are species-specific and have their roots in the biological nature of man. Perception and conceptualization may be essentially the same in all cultures. The theoretical

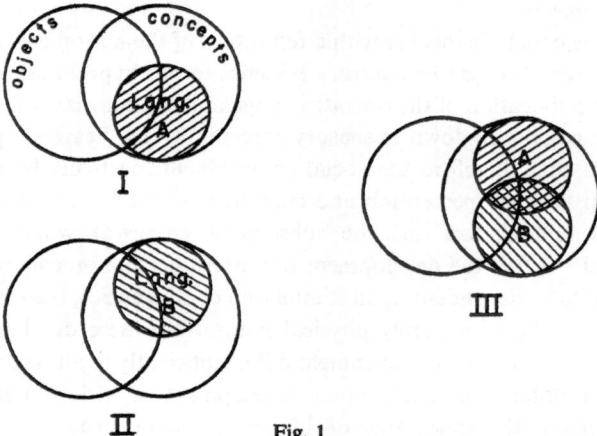

Fig. 1

relationship between language, concepts, and the physical world (objects) is shown in Figure 1. Concepts are not coterminous with objects; language (or words) are not coterminous with either. Language ordinarily only reflects *a few* aspects of the speaker's conceptualization but it *can* be extended *ad hoc* to any area in the realm of concepts (by creating new terms in science, for instance); in a sense, natural languages "spot-sample" the realm of concepts which itself is potentially universal or accessible to *homo sapiens* in all cultures. Diagrams I and II in Figure 1 show how two languages may draw different samples from the realm of concepts and diagram III shows how a comparison of such two languages may produce the untranslatability that puzzled Whorf so much.

The semantic differences between languages may be much less dramatic than it seemed at first. It is conceivable that, and probable that, all languages roughly deal always and universally with similar phenomena and relationships, say color, number, time, and certain objects, leaving a

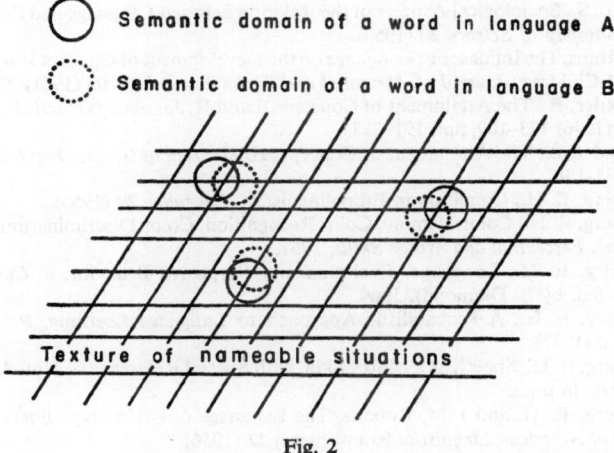

Fig. 2

great many other relationships that *might* theoretically be "talked about" unnamed. Figure 2 shows how the semantic shifts between languages might be relatively small considering how few of the many possible phenomena or relationships (cross-points in the figure) are regularly left unnamed and also the possible large distance between named phenomena (between the three ring configurations), in comparison with the close

53

coincidence of what is to be named (that is distance between matching solid and punctuated circles).

In tasks where language is the only possible (or most easily accessible) "information carrier", language structure may affect cognitive processes. Thus language is capable of widening the cognitive horizon, but its absence does not limit the individual to a "sub-human" cognitive state.

Harvard Medical School, Cambridge, Massachusetts

REFERENCES

1. Black, Max., Linguistic Relativity: the Views of Benjamin Lee Whorf, *Philosophical Review* **68** (1959).
2. Brown, R. W., *Words and Things*, 1957.
3. Brown, R. W. and E. H. Lenneberg, A Study in Language and Cognition, *Journal of Abnormal and Social Psychology* **49** (1954).
4. Bruner, Goodnow and Austin, *A Study of Thinking*, 1956.
5. Feuer, L. S., Sociological Aspects of the Relation between Language and Psychology, *Philosophy of Science* **20** (1953).
6. Furth, Hans, The Influence of Language on the Development of Concept Formation in Deaf Children, *Journal of Abnormal and Social Psychology* **63** (1961) 386–389.
7. Heidbreder, B., The Attainment of Concepts, I and II, *Journal of Genetic Psychology*, **35** (1946) 173–189, and 191–223.
8. Heidbreder, B., The Attainment of Concepts III, *Journal of Genetic Psychology* **24** (1947) 93–138.
9. Lenneberg, E. H., Cognition in Ethnolinguistics, *Language* **29** (1953).
10. Lenneberg, E. H., Color Naming, Color Recognition, Color Discrimination: a Reappraisal, *Perceptual and Motor Skills*, 1961.
11. Lenneberg, E. H., Language, Evolution and Purposive Behavior, in *Culture in History* (ed. by S. Diamond), 1960.
12. Lenneberg, E. H., A Probabilistic Approach to Language Learning, *Behavioral Science* **2** (1957).
13. Lenneberg, E. H., Speech as a Motor Skill, with Special Reference to Non-Aphasic Disorders. In press.
14. Lenneberg, E. H. and J. M. Roberts, The Language of Experience, *International Journal of American Linguistics* (supplement) **22** (1956).
15. Olerory, Pierre, *Recherches sur le Développement Mental des Sourdes-Muets*, 1957.
16. Quine, W. van O., *Word and Object*, 1960.
17. Teuber, H. L., *Perception, Handbook of Physiology: Neurophysiology*, 1961, vol. 3, ch. 4.
18. Whorf, B. L., *Language, Thought and Reality* (ed. by J. B. Carroll), 1956.
19. Chomsky, A. N., *Syntactic Structures*, 1957.

3

LASZLO TISZA

THE LOGICAL STRUCTURE OF PHYSICS *

Presented December 14, 1961

I. THE LOGIC OF CONCEPT FORMATION

Crystallizations of genuinely new concepts that are not combinations of concepts already established often mark turning points in the evolution of physics. In retrospect it is possible to view such processes of concept formation as interplays of two types of procedures. The first is what might be called *conceptual differentiation*. It arises as concepts distilled from past experience are brought to bear on experimental situations that are either novel, or at least more detailed and discriminating than those of the past. In such cases the original concept is usually modified and subdivided leading to a bifurcation or differentiation. An example is the splitting of the undifferentiated concept of *calor* (heat) into *temperature* and *caloric*, and the further subdivision of the latter into *energy, entropy* and *heat quantity*.

Counteracting the multiplication of concepts by differentiation is the occasional discovery of new fundamental concepts that enable us to represent concepts, that had been believed to be primitives, as constructions. An example is the electromagnetic field that made the unification of the theories of electricity, magnetism and optics possible. We note that whenever we speak of *concepts* we should include the *principles* formalizing their use. Thus the electromagnetic field plays its unifying role only if defined in terms of Maxwell's equations, including the displacement current.

The proper functioning of conceptual differentiation and integration is beset by characteristic difficulties. Differentiation may occur in a large number of small steps and the realization that a radical conceptual change has taken place may be long delayed. The difficulty is usually aggravated

---

* A more detailed presentation is found in 'The Conceptual Structure of Physics', *Reviews of Modern Physics*, January 1963. We refer to this paper as *loc. cit.*

by semantic confusion as concepts which are actually used in different senses in different contexts are still designated by the same term. It was recently pointed out by Karl Menger [1] that this difficulty arises even in contemporary mathematics. According to Menger "... mathematico-scientific methodology is in need – in fact, in urgent need – of a *separator* or a *prism* resolving conceptual mixtures into the spectra of their meanings". It would seem that the need for such a prism in physics is still more pressing.

The difficulties associated with conceptual integration are even greater. A limited field of experience can usually be represented in terms of several formal theories. The comparative merit of these formalisms often becomes apparent only by confrontation with phenomena belonging to quite remote specialties. Such a confrontation is always difficult, but it is becoming more so with the widening gap between specialties.

The main purpose of this paper is to outline a method of logical differentiation and integration that should enable us to proceed with the formation of concepts in a more controlled and systematic fashion. In particular, this method ought to supplement the subjectivity of historical evolution with a more objective appraisal of logical connections. In this method of analysis deductive systems play an essential role, a statement presumably surprising in view of the curious ambiguity surrounding the appreciation of deductive systems in physics.

The deductive method is quite prominent in the early development of mechanics by Galileo and Newton. Also the great systems of 19th century physics can be considered as somewhat loosely organized deductive systems. However more recently the use of deductive systems has declined. The impression seems to prevail that deductive systems are rigid, narrow and lead to the ossification of theories by lending authority to obsolescence. There is, of course, no denying that this description of the deductive method is often entirely accurate. However, the potentialities of this method are in striking contrast with its actual uses in the past. If properly handled, deductive systems can become instruments of radical innovation even while safeguarding the genuine achievements of the past. In essence two ideas are responsible for this change in outlook. In order to avoid rigidity, the postulational basis, or briefly, the basis of deductive systems has to be considered as hypothetical rather than an expression of self-evident truth. Therefore the basis is subject to change if this leads to an improvement of the system in accounting for experimental facts. This feedback from system to basis leads to the dynamic, evolutionary adjustment of the system to a widening range of experience. Instead of assuming that deductive systems have to be perfect in order not to collapse, the present method of analysis deals with imperfect systems. In fact, one of the tasks of the method is to locate the eliminate imperfections. In view of this situation, the method can be aptly called *dynamic logical analysis*, or the *dynamics of deductive systems*.

While the idea of the hypothetical nature of deductive systems is, in principle, well known, it has not yet led to practical consequences for structuring the theories of physics. To achieve such applications we have to overcome the difficulties stemming from the narrowness of deductive systems. This narrowness is an irreducible fact of logic, and the scope of systems is usually further narrowed as the rigor and completeness of the formulations is improved. The cure suggested for this difficulty is the simultaneous use of many deductive systems. Whereas traditionally a deductive system was usually the entire universe of discourse, according to the present point of view deductive systems are themselves conceptual elements and their interrelations, particularly the conditions of mutual consistency, become the main objects of study. In many important cases these relations exhibit typical patterns some of which we shall consider in the next section. An application of these ideas to the conceptual problems of quantum mechanics will be considered in Section III.

II. LOGICAL DIFFERENTIATION AND INTEGRATION

It is an important insight of mathematical logic that each deductive system contains primitive concepts that cannot be defined by conventional means, but are implicitly defined by the deductive system as a whole. This "definition by postulation" of primitives, so important in modern mathematics, turns out to be valuable also in physics. While recognizing the basic identity of the method in both disciplines one should also take note of a difference stemming from the fact that the deductive systems of physics are always supplemented by rules of correspondence which establish a link with experience. Mainly as a consequence of this connection, most concepts of physics evoke in the mind images or models, for which the postulates are presumably true statements. Although this

57

intuitive aspect of the abstract concepts of physics is important for many reasons the relations between experiment, intuitive models and deductive systems are intricate, not unique, and hard to state in precise terms. The method of definition by postulation is an effective tool for clarifying the semantic ambiguities that arise out of this situation.

The fact that primitive concepts assume a precise meaning only relative to the deductive system into which they are incorporated, can be aptly referred to as the *relativity of concepts*.

Another interesting aspect of the present analysis is the substantial increase in the number of deductive systems. It appears that most of the basic disciplines, say classical mechanics, thermodynamics, quantum mechanics, are too ramified and diverse to be compressed within the confines of a rigorously built deductive system. This point has, of course, been advanced as an argument against the use of deductive systems in physics. Instead, we propose to overcome this difficulty by constructing a cluster of deductive systems, even for representing one basic discipline.

Since, as we have pointed out, each deductive system implicitly defines its own set of primitive concepts, the increase in the number of systems is tantamount to an increase in the number of fundamental concepts that are defined relative to a particular context specified in terms of a *deductive system*. For this reason, we designate the procedure leading to a collection of deductive systems as *logical differentiation*.

The conceptual wealth brought about by the method of logical differentiation provides us with all the freedom we may need to adjust our deductive systems to a variety of experimental situations.

However, if the analysis should stop at this point, one might feel that the freedom in the formation of deductive systems has been carried too far, since it seems to lead to a proliferation of basic concepts. Actually, however, this process is held in check by the second stage of our procedure, the *logical integration* of deductive systems. The conditions of mutual consistency of empirically interpreted deductive systems have empirical and logical aspects. A study of these problems should be classified as part of *logical empiricism*, although the present discussion goes beyond the traditional scope of this discipline. The possibility of a concise discussion of the general problem of mutual consistency hinges on the circumstance that the formal logical aspects of these compatibility conditions can be recognized within a model of transparent simplicity.

The same formal conditions can be given different concrete interpretations and used to handle most intricate situations.

The model in question is the practical (physical) geometry of a spherical surface, in other words geodesy. The physicist-geometer can avail himself of two abstract geometrical systems, to wit spherical, and plane geometry. From the point of view of the abstract mathematician the two systems are different and a discourse has to be completed in one *or* the other system. The consistent use of the two systems for the description of the real, physical sphere depends on a number of circumstances.

In the first place, we establish a hierarchy between the two systems: we stipulate that in case of conflict spherical geometry is the *dominant* system, and plane geometry is called either *derivative, subordinate,* or *degenerate.*

We shall refer to the relation of these systems as supplementary *relation,* or, briefly, *supplementarity.*

The second point is a clarification of what constitutes a "conflict" between the systems. In the mathematical sense a conflict always exists; however, in physical geometry we dismiss it as empirically irrelevant if it falls below the finite accuracy of the measurements which the theory is supposed to describe. The accuracy of measurement cannot be improved over the diffuseness set by thermal noise.

Somewhat more generally we proceed to formulate a principle to which the rules of coordination have to conform. Experiment provides us with the values of continuous variables within a certain accuracy. However, the analysis of mathematical physics would be extremely cumbersome and lose much of its precision if the finite width of continuous parameters corresponding to empirical quantities were to be observed in every step. In actual practice, the segments of finite widths are sharpened into definite points of the continuum for the purposes of the analysis. However, the results obtained have no physical meaning unless they are sufficiently insensitive to the actual unsharpness of the continuous parameters. Mathematical solutions satisfying this requirement will be called *regular.* Solutions that change their essential features when the fictitious sharpening is given up will be referred to as *singular.* They are devoid of physical meaning except as possible stepping stones to more realistic results. The requirement that mathematical solutions be regular in order to have a physical meaning will prove to be of great importance and deserves to be called a principle, the *principle of regularity*.

The second point concerns the question of *scale*. Spherical geometry is characterized by an absolute unit of length R, the radius of the sphere. If $R \to \infty$, the sphere degenerates into the plane, the dominant into the supplementary system. Let the linear dimension of a geometrical figure be a, then the parameter a/R is appropriate to express the deviation of spherical from plane geometry. It becomes meaningful to say that for small values of the parameter ($a/R \approx \epsilon \ll 1$,) where ϵ is the unsharpness of the parameter because of the inaccuracy of the measurement or because of noise, the subordinate system is asymptotically valid. Otherwise we have to fall back on the dominant system.

It is interesting to note that within plane geometry there is no absolute unit and it is meaningless to speak about small or large figures. All theorems are invariant under a transformation consisting of a change of the unit of length. We say also that plane geometry is *scale invariant*, or, it contains the theorem of similarity.

It is well known that deductive systems provide the implicit definitions of their primitive concepts. It is easily seen that the limiting process leading from the dominant to the derivative system profoundly modifies the conceptual structure of the deductive systems concerned. While the conceptual structure of spherical geometries, which differ only in the numerical value of the fundamental radius R, is identical, there is a discontinuous change in the conceptual structure as R becomes infinite. We are confronted here with the logical analog of nonuniform convergence. Paradoxically, the deceptive simplicity of the model may provide the greatest hurdle to an appreciation of the relation between the two systems. It seems to be almost as hard to recapture a feeling of puzzlement in connection with a paradox completely resolved, as to get rid of a feeling of discomfort when faced with essentially the same paradox in a novel context. Yet for those who believed that the earth was flat, the relativization of "up" and "down" and the existence of antipodes on the sphere was no less puzzling than the relativization of time for someone thinking in terms of classical kinematics. The logical and even the mathematical connection between the two situations are very close. Moreover, if one believes that the sum of the angles of triangles has to be π, then one will conclude that the concept of triangle "breaks down" on the sphere.

One of the conceptual problems to be solved in connection with the individual dominant systems is that of invariance. The transition from the subordinate to the dominant system marks the breakdown of scale invariance. At the same time, however, the dominant systems is characterized by an invariance of a novel type. In spherical geometry we have invariance of curvature.

In the course of historical evolution the dominant system has to be established by generalizing the subordinate system that reached what we shall call a *marginal breakdown*. In general, this is a task of great difficulty. While the converse procedure of arriving through a limiting process from the dominant to the subordinate system is a great deal more straightforward, even this transition is not without pitfalls.

Just as the sphere does not contain the infinite plane, so spherical geometry does not contain plane geometry. While the limiting process $R \to \infty$ transforms every theorem of spherical geometry into one of plane geometry, the converse is not true. Thus the important laws of similarity of triangles and other figures have no counterpart in spherical geometry. The possibility of deriving such theorems is a new discovery, possible only in plane geometry. This situation is quite similar for other pairs of dominant-supplementary theories.

We shall say that the inconsistency between the subordinate and the dominant system is *under control*, or that the *inconsistency* is controlled, if the range of validity of the subordinate system is known, and within these limits the compatibility with the dominant system is ensured.

Figure 1 contains a simplified, symbolic representation of the integrated structure of systems that results from the application of the principles discussed in this section. Although this structure, briefly, the *block diagram*, is still incomplete and tentative, it is adequate to provide the point of departure for our discussion. We have to remember that our method is set up to eliminate the shortcomings of the initially posited structure rather than to perpetuate them.

Maybe, the most significant aspect of the block diagram is the pluralistic character of the structure exhibited. Moreover, CMP [2] appears as derivative to several systems and dominant to none. This is in complete contrast with the logical structure hypothesized in classical physics in which CMP should have been dominant to all the other deductive systems of physics.

For even a sketchy discussion of the block diagram we have to refer to *loc. cit.* However in the next section we shall summarize some of the conclusions concerning the conceptual problems of quantum mechanics.

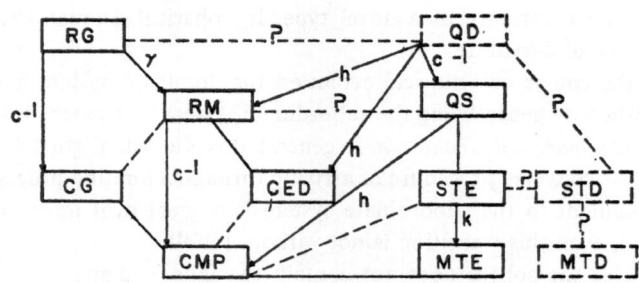

————COMPATABILITY————CONTROLLED INCONSISTENCY,
→ SUPPLEMENTARY RELATION ——— ? ——— INCOMPLETELY
UNDERSTOOD RELATION

TENTATIVE STRUCTURE OF PHYSICS

Fig. 1

c^{-1} inverse speed of light; h Planck's constant; k Boltzmann's constant; γ gravitational constant.

Systems in broken rectangles are essentially incomplete.

Symbols of systems:

CMP	Classical mechanics of particles	TD	Thermodynamics
CMP(μ)	Classical mechanics of particles (microscopic interpretation of particles)	MTE	Macroscopic thermodynamics of equilibrium
		STE	Statistical thermodynamics of equilibrium
PhCM	Phenomenological classical mechanics	MTD	Macroscopic thermodynamics
CG	Classical gravitational theory	STD	Statistical Thermodynamics
CED	Classical Electrodynamics	QM	Quantum mechanics
RM	Relativistic Mechanics	QS	Quasi-static quantum mechanics
RG	Relativistic Gravitational-Theory	QD	Quantum dynamics

III. THE TWO PRINCIPLES OF CAUSALITY

The empirical basis of the physical sciences is provided by two major disciplines, astronomy and chemistry, the formalization of which gives rise to CMP and TD. The former adequately describes the motion of well separated macroscopic objects, whereas TD deals with intrinsic structural properties. Quantitative connections are much more easily observable in astronomy than in chemistry. It was presumably unavoidable that CMP should be the first discipline to develop into a powerful and versatile formalism, while TD was late to start, slow to develop and is even now in an "underdeveloped" state measured in terms of its potentialities. Since the traditional unitary attitude, fostered by the unification of celestial and terrestial mechanics, could admit only one fundamental deductive system, it was taken for granted that, say TD would eventually be reduced to CMP. In contrast, according to the pluralistic view TD can be admitted as a semi-autonomous system with its *sui generis* primitive concepts, even if some of these are evidently not reducible in a rigorous sense to CMP.[3, 4] Thus TD and CMP implicitly define two conceptions of matter which are strictly speaking in conflict with each other. This inconsistency indicated by a broken line in the block diagram is brought under control by QM.[5] The application of the present method allow us to conclude that the conceptual structure of QM is much closer to TD than to CMP. It is for historical reasons that this fact did not become manifest thus far. QM has been insufficiently differentiated from CMP and insufficiently integrated with TD. This was made practically unavoidable by the incomplete development of TD.[6]

We turn now to a short summary of some of the features that make up the conceptual contrast, or rather complementarity of CMP and TD.

Instead of building up systems out of mass points, as in CMP, in TD one builds them out of "cells" that cover a certain region in space. We refer to the cells as *simple systems*, or *subsystems*, and to a collection of cells as a *composite system*. The independent variables of the theory are additive conserved quantities, briefly, additive invariants X_1, X_2, ... Composite systems are specified by sets of the X_1 in the subsystems of the composite system.

Explicitly, the X_1 are the volume V, the internal energy U, and the mole numbers of the independent chemical components N_1, N_2, ... The characteristic coupling of thermodynamic systems comes about through the exchange of energy and of other additive invariants.

We turn now to the discussion of processes. In CMP all processes are alike, inasmuch as all of them are trajectories in phase space. In TD we have processes of different types. An important one is the transfer of a quantity X from one subsystem to another during an arbitrary, fixed time interval. The transfer quantity corresponding to the energy, with all of the other X's fixed, is called *heat*.

Walls are the boundaries separating two systems which completely prevent the passage of one or more additive invariants, but permit free passage to all others. The restrictions on a system imposed by its walls are referred to as *constraints*.

The concept of thermodynamic process is easily extended to chemical reactions. In this case the molecular potential barriers stabilizing chemical species play the role of walls.

We call *thermodynamic operations* any imposition or relaxation of a constraint through the uniting or subdividing of systems, or the alternating of the type of any of its walls.

The distinction of thermodynamic operations and processes is a characteristic feature of the present approach. The admission of thermodynamic operations into the framework of the theory allows one to account for experimental procedures that have no place within CMP because they correspond to transitions from one phase space to another.[7] Thus thermodynamics allows us to account formally for manipulations on thermodynamic systems. In particular, such manipulations may be performed on systems in the course of a measurement. Yet the insistence on operations as contrasted with processes does not necessarily require the intervention of an experimentalist, and does not introduce more subjective elements into the theory than are warranted by the nature of the situation considered. Thus in the case of chemical reactions and phase transitions, supplying catalysts or inhibitors that speed up or slow down the rates of reactions, is the corresponding thermodynamic operation. In an automatic chemical plant, and even more in biochemical processes, the "operations" become effective without the interference of an experimentalist and it may be more adequate to speak of parametric processes which *control* or *govern* other processes. This is quite in line with the use of these terms in cybernetics.[8]

In the present context the approach to equilibrium is described as follows. In a composite system in equilibrium the relaxation of an internal constraint (a thermodynamic operation) triggers a process leading to a new equilibrium, in which the partition of the additive invariants is uniquely determined by the nature of the system and by the existing constraints. Equilibrium is uninfluenced by past history, specified e.g. in terms of constraints that had been relaxed previously. This important result has been designated [3] as the *principle of thermostatic determinism*, suggesting a parallel with the *principle of mechanical determinism*. The latter has, of course, a temporal character and the connection between the two principles is traditionally established by reducing thermostatics to mechanical determinism. However it was pointed out *loc. cit.* that the sequences of thermodynamic operations and processes can be iterated to form composite processes. Due to the fact that the principle of thermostatic determinism assures us of the uniqueness of the final state of each step, the entire composite process assumes a deterministic character, provided that the operations are present by some automatic device. Alternatively, if the operations are governed by the terminal states through feedback, the composite process becomes goal-directed.

The temporal determinism of mechanical processes depends on the *exhaustive specification* of the initial state of the system involved. If, as usual, the system is coupled to its environment, then the initial state of the environment has to be also completely specified; in practice this is an impossible task. Therefore mechanics accounts for the deterministic behavior of systems only if these are effectively decoupled from their environment. Evidently the case of planetary systems plays a uniquely favorable role in this respect. Being sheltered from the environment by vast reaches of space, the specification of its initial state can be practically accomplished with the accuracy required for macromechanical predictions extending over long intervals of time.

The impressive success of this procedure, together with the possibility of generalization to the differential equations of CED and QM, lead to the conviction that *all* causal chains in nature are conditioned by an exhaustive specification of the initial state. This conviction seems to be still quite widespread, although it is untenable from the point of view of experimental evidence. It is a fundamental fact in all experimental sciences that there is a selectivity of the specification of initial states that distinguishes between *essential* and *irrelevant* factors. It is this *selectivity* of specification that enables us to experiment at all, and to utilize the results of past experience under new conditions that are never exactly duplicated.

65

This fact becomes most impressive in living organisms: the seed of a plant develops into a well-determined organism, provided only that the environment to which it is coupled satisfies some very weak requirements concerning the range of temperature, pressure, and chemical potentials.

This principle of causality based on selective specification finds a theoretical foundation in thermodynamics which leads to the temporal determinism of composite processes just described. Thus we are led to *discern two principles of causality.* While the *first principle is based on exhaustive specification of the initial state,* the *second depends on selective specification.* The two principles are complementary: The first is *incompatible* with stochastic laws [9]; the second requires such laws. Thus there is a built-in conflict between the theoretical and the experimental aspects of Newtonian natural philosophy, since they depend on the first and the second principles of causality. Critical philosophers such as George Berkeley and David Hume sensed these difficulties, but they were unable to resolve them. In contrast, physicists were too busy developing the rich implications of both methods to be slowed down by conceptual scruples that were not yet ripe for solution.

The foregoing considerations are considerably strengthened and sharpened when extended to quantum mechanics. I will have to confine myself to a few remarks and refer for further details to *loc. cit.* As pointed out above, in TD we consider transitions between equilibrium states. This ties in naturally with the transitions between quantum states in QM. Particularly interesting are the *pure states* which represent *potential* forms for the existence of objects. If such a state *actually* exists, we say the state is occupied, or realized. *The realizations of the same pure state in different regions of space time form a class of absolutely identical objects.*

The time-independent Schrödinger equation allows one to compute all pure states that can be constructed from a given assembly of particles. Thus, in principle, all of the pure states of structures built from a given set of particles can be computed on theoretical grounds without any recourse to experiment! Whenever this calculation has been actually carried out, the results have been verified by experiment to a high accuracy. There are undoubtedly huge classes of cases in which the same agreement could be achieved, provided that one overcomes the mathematical difficulties that increase enormously with the complexity of the systems.

Although the calculation of the properties of pure states by exclusively theoretical means is of the greatest philosophical interest, it is no less important that the taxonomical task of listing and specifying the set of pure states can be undertaken by judicious combination of theory and experiment. It can be shown that the *"retrodiction" of the discrete quantum numbers specifying pure states from the statistical evaluation of actual signals can be carried out in principle, and very often in practice, with absolute certainty.*

We may sum up the situation by saying that QS contains rigorous results of a taxonomical character concerning the existence of classes and their properties which are entirely foreign to classical physics. Remembering from Section II that dominant systems have their characteristic invariance property, we say that the absolute identity of any realization of the same pure state is the dominant invariance of QS, to be referred to as *morphic invariance*. It is the morphic invariance that brings it about that complex systems of well-defined identity can be built in nature through the mechanism of random processes.

A preliminary discussion (*loc. cit.*) shows that the combination of the concepts of morphic invariance with the two principles of causality provides a theoretical framework for the description of the functioning of organisms.

IV. PHILOSOPHICAL REFLECTIONS

The method of analysis of this paper is, I believe in agreement with the spirit and the underlying goals of the philosophical school that is alternately called analytic, positivist or empirical. I hope not to be too far off the mark in using these terms as synonyms. The concepts of logical differentiation and integration arose within the examination of the language of physics, which is a legitimate pursuit in this school of thought.

A point of agreement is the close attention paid to empirical requirements; more specifically, the approach to quantum physics as described *loc. cit.* is quite in line with the positivist program of developing the theory of the structure of matter in close parallel with experiment. The principle of regularity (Section II) is a methodological rule that serves as a safeguard against taking theoretical systems more seriously than their empirical support warrants.

At the same time, the present paper and its implications are at variance with some of the current practices of positivism. The clarification of these issues can be achieved as a by-product of the present analyses.

Let us consider first, the so-called *operational character* of the theoretical concepts. It goes without saying that the deductive systems of physics have to have a range of concordance with experience.

Yet, there is a growing awareness of the fact that there is something wrong, or at least lopsided about the great emphasis on "operationalism" which for several decades has dominated empirical philosophy.

We could do hardly better than quote from P. W. Bridgman's [10] penetrating reappraisal of the early ideas of operationalism. "If I were writing the Logic today I would change the emphasis so as to try to avoid what I regard as the one most serious misunderstanding. That is, I would emphasize more that the operations in terms of which a physical concept receives its meaning need not be, and as a matter of fact are not, exclusively the physical operations of the laboratory. The mistaken idea that the operations have to be physical or instrumental, combined with the dictum on page 5, 'The concept is synonymous with the corresponding set of operations', has in some cases led to disastrous misunderstanding. If I were writing again I would try to emphasize more the importance of the mental or paper-and-pencil operations. Among the very most important of the mental operations are the verbal operations. These play a much greater role than I realized at the time..."

The postulational definitions of the present approach are of course "paper-and-pencil" operations. Apart from being of equal importance with the instrumental operations, the two are related to each other in a well-structured manner. In the absence of precise postulational definitions the meaning of the instrumental operations cannot even be properly evaluated. The operational point of view considered as a sole criterion would always favor the low-level abstractions. However, the most spectacular advances in theory are connected with the discovery of abstractions of a very high level.

The constructive aspect of operationalism is appreciated particularly in the process of logical integration. If we have two theoretical systems, both of them confirmed by experiment, but inconsistent with each other, we usually find that the inconsistencies are produced by nonoperational assumptions that can be dropped without loosing the measure of experimental agreement already achieved. In fact, subsequent developments often result in an extraordinary expansion of the range of concordance.

This is what happened as CMP was generalized into the dominant RM. Einstein's operational analysis of the concept of simultaneity was a means to an end, it made the integration of CMP and CED possible. The concept of absolute simultaneity was abandoned because it blocked integration. The fact that it was not operational showed that this change in the foundation could be undertaken without damage.

However, "logical integration" has been, thus far, not in the vocabulary and the *purpose* of operationalism was not sufficiently appreciated. Erroneously it was taken for an end in itself, or, an ironical perversion of its real role, a means to forestall another "catastrophe" that would force us to re-examine our basic assumptions.

Although an occasional reappraisal of the foundations may seem a not entirely welcome interruption of the routine of research work, it is now recognized to be the unavoidable price to be paid for the continued use of new high-level abstractions. We again quote Bridgman: "To me now it seems incomprehensible that I should ever have thought it within my powers, or within the powers of the human race for that matter, to analyze so thoroughly the functioning of our thinking apparatus that I could confidently expect to exhaust the subject and eliminate the possibility of a bright new idea against which I would be defenseless."

We turn now to another difficulty of positivism, namely its predominantly restrictive character which bids us to dismiss most problems of traditional philosophy as "meaningless". Granted that the traditional methods are lacking in precision, we are faced with the unhappy choice of dealing with significant problems in an unsatisfactory manner or bringing to bear a precise method on problems the wider import of which is not immediately apparent.

There are at least two ways of breaking this deadlock. One could make a case for applying less rigid standards in appraising intuitive methods, or to extend the scope of analytic philosophy to more significant problems. It is only the second path that will be followed up at this point.

In order to understand the origin of the restrictiveness of positivism, we have to go back to the beginnings of mechanistic physics. With the brilliant success of Newtonian mechanics, it seemed tempting to brush away the complex and eroded conceptual system of scholastic philosophy. The proposition of temporarily restricting oneself to the conceptual framework of the new mechanics seems entirely sound, even in retrospect. However, the contention that this conceptual framework would be satisfactory at all times, was unwarranted, and turned out to be actually incorrect. During the mechanistic era it became customary to dismiss types of questions that did not fit into mechanistic systems as unscientific. As the crisis of classical physics revealed the limitation of the mechanistic conceptual scheme, the first inference was that the range of *legitimate scientific questions is even further limited*, since not even the mechanistic questions are admissible.

The pluralistic character of the present approach brings two new elements into this picture. In the first place, each deductive system implies a characteristic set of precise questions. The number of interesting questions that become "meaningful" is particularly extensive in thermodynamics and quantum mechanics.

It is often stated that the concept of object breaks down in quantum mechanics. Actually, however, the opposite is true. As outlined in Section IV, of *loc. cit.* for the first time in QM we are in a position to give a formal representation of an object with many subtle ramifications and we can now resolve the Eleatic paradoxes stemming from the apparent inconsistency of identity and change. It seems that the new object concept is flexible enough to include living organisms that are entirely outside the mechanistic scheme.

The extension of meaningful conceptual problems in the present context proceeds in still another dimension. Not only do we have the concepts within each deductive system, but the deductive systems themselves are conceptual entities of distinct individual characteristics related to each other in quite specific fashion. These entities are of a logical type that is markedly different from that of the primitive concepts within the deductive systems.

Massachusetts Institute of Technology, Department of Physics, Cambridge, Massachusetts

NOTES

1. Karl Menger, A counterpart of Occam's razor in pure and applied mathematics ontological uses, *Synthese* **12** (1960) 415.
2. Henceforth we shall designate the deductive systems by the abbreviations listed in the caption of Fig. 1.

3. L. Tisza, The Thermodynamics of phase equilibrium, *Ann. Phys.* **13** (1961) 1.
4. L. Tisza, and P. M. Quay, S. J., Statistical Thermodynamics of Equilibrium. To be published.
5. In the block diagram QM is differentiated into QS and QD. For a discussion of these finer points we refer to *loc. cit.*
6. It is interesting to compare this situation with another one closely analogous from the logical point of view. The inconsistency of CMP and CED was brought under control by RM. The space-time structure of this theory is consistent with that of CED, but distinct from CMP and dominant to it. However, in this case the common understanding of the situation is more satisfactory, presumably because early in this century CED was in a more advanced state of development than TD.
7. The term "mechanics" is generally used in an ambiguous fashion either to denote analytical dynamics or any of a variety of more phenomenological disciplines. We refer to Section IV B of *loc. cit.* where logical differentiation is used to resolve the conceptual spectrum. In the above discussion CMP is taken in a narrow sense as analytical dynamics.
8. N. Wiener, *Cybernetics*, 2nd edition, New York, J. Wiley and Sons, Inc., 1961.
9. These are probabilistic laws connecting the states of a system at different times with each other.
10. P. W. Bridgman, P. W. Bridgman's 'The Logic of Modern Physics' after Thirty Years, *Daedalus* **88** (1959) 518.

DISCUSSION

PHILIPP G. FRANK, LASZLO TISZA, ROBERT MAYBURY,

FRITZ STECKERL, WILLIAM WALLACE, ARMAND SIEGEL,

ABNER SHIMONY

Professor Frank: I will only say something on the last part, on the question of operationalism here. As you know, Bridgman took his operationalism from Einstein, but there was from the beginning a certain discrepancy. Bridgman's original idea probably was that in any critical discourse every word should be defined by an operational meaning. And Einstein at the beginning probably agreed with this. But later, I remember the moment when Mr. Bridgman asked me to go for him to interview Einstein, and to ask him flatly, bluntly: "Did you change your opinion, or do you still believe that every concept in a scientific discourse should have an operational meaning?" And Einstein said: "You see, I wouldn't say this. I would not say that every word must have an operational meaning, but it should be required that *in the consequences of every word used there must be an operational meaning.*" Also I may add on my part that sometimes the operational meaning isn't explicit, but implicit in the language. At any rate, whatever the intermediate non-operational steps may be, at the end there should be an operational meaning.

Professor Tisza: This is a very graphic statement of the problem and I agree, of course, with the revised formulation of operationalism. I believe that this agreement is general among physicists and, according to the above quotations, this includes also Bridgman in his later years. The problem at present is to catch up with the wide repercussions of the earlier exaggerated statements. In order to achieve this it may be helpful to point out that the early versions of operationalism stemmed from a very natural reaction to the stunning breakdown of the classical concepts. These were so firmly entrenched in tradition as to be attributed to the immutable forms of our thinking apparatus. The strict adherence to empiricism and distrust of a crippling conceptualization were the proper emergency measures in this situation. The development of the physical sciences during the last half century are indications of the success of the new departure, the boldness of which it is hard to appreciate at present. However, precisely the success of the last decades bids us to reconsider some of the attitudes borne out of the emergency. The situation seems more than ripe for a new conceptualization firmly rooted in the rich soil of the new experience, and freed from obsolete patterns of thought.

Professor Maybury: You outlined, in a sense, a hierarchy of determinisms or causalities, or better yet "unfoldings" of states: one, the mechanical predictions of a path; and then second, that which we have in a thermodynamics. I wonder if you could just go back over that again, especially in what sense you thought that the thermodynamic unfolding of states is a more precise or a more adequate causality than that which we usually associate with mechanical causality.

Professor Tisza: I would rather say that there are two complementary principles of causality. The first is the mechanical one, it depends on the exhaustive specification of the initial state and is adequate primarily in celestial mechanics. The second, depending on the selectivity of determining factors is indispensible for the understanding of chemical and biological phenomena, the main features of which exhibit quite stable causal features even though the systems are strongly coupled to their surroundings. This is incomprehensible in terms of the first principle of causality, according to which the slightest variation in the surroundings produces substantial changes in the course of events. Generally speaking the two principles of causality have to be applied jointly.

Dr. Steckerl: In the long run do you think you will have a deductive system, one royal one – you know, a dominant one, spreading out over all possible statements; or may there be independent deductive systems, and which one you choose will just depend on the kind of question you ask?

Professor Tisza: Maybe, but I do not know. It is an important aspect of the present method that, in contrast to the traditions of speculative philosophy, we do not have to, in fact we should not commit ourselves regarding such deep problems. The important matter is the method that allows us to improve in a systematic fashion the logical structure of the existing theories. I believe that this method of analysis, in conjunction with new experiments, will lead to a reasonably unified logical structure. Strong beliefs concerning the specific features of this structure might interfere with the free functioning of the method.

Rev. Wallace, O.P.: In your diagram the field theories show a type of duality and seem to converge to relativistic gravitational theory and to quantum dynamics. Have you thought about quantizing the gravitational field? Professor Wheeler at Princeton has worked on this problem. He seems to provide an interesting synthesis of the two field theories.

Professor Tisza: I am not familiar with the attempts at quantizing the gravitational field. This is a fundamental problem that involves the integration of quantum theory with general relativity. It is natural that pioneering work should be done on it. However, I believe that a satisfactory solution may come only after the integration of quantum field theory with special relativity has been accomplished with conceptual clarity. In spite of impressive progress along this line, this aim has not yet been attained. However, your remark on the duality of the two field theories reminds me of a quite specific duality that exists between the geometrical formalism of RG and of STE. A short statement of this duality is given in footnote 4 of reference 3. It is suggestive that a geometrical duality should exist between the two field theories that deal with non-specific gravitational and with specific chemical interactions of matter.

Professor Siegel: I think I'm quoting Tisza when I refer to the kind of determinism in which two chemical reagents give rise to a third, and some kinds of processes involving biological determinism – you once called these morphic determinisms. Is this a kind of determinism whereby a given set of forms determines another set of forms?

Professor Tisza: The two concepts are not entirely the same. A few years ago I advanced the term "morphic determinism". Although I have not changed my mind on the ideas contained in this unpublished manuscript, I prefer now a different terminology. First we have "morphic invariance". This is implicit in QS, the part of quantum mechanics centered around the time-independent Schrödinger equation. A most succinct statement of this principle is as follows: The set of possible pure states of stable atomic systems is countable since it is specified by discrete quantum numbers. Moreover, different realizations of the same pure state are absolutely identical replicas of each other. As outlined *loc. cit.*, morphic invariance is largely responsible for the fact that there *is* an atomic physics, that precise inferences about atomic systems can be drawn from feasible experiments. Temporal concepts play only a marginal role

74

in these considerations. A full quantum dynamics is not yet available for temporal problems. However, the two principles of causality account qualitatively for the observed temporal causal chains.

Professor Siegel: I didn't mean to use your old terminology to commit you. However, I wish to emphasize the looseness of the temporal determinism you emphasize so much. If you seek support from cybernetics, this is a very strongly stochastic probabilistic theory. And from the point of view which Einstein emphasized in his criticisms of quantum mechanics, he would have said that this was certainly a powerful theory, but it was an incomplete theory. That is, although quantum mechanics might be dominant to classical mechanics, if quantum mechanics by virtue of the stochastic element is incomplete, and if there is some more complete theory behind it, then it is not dominant and the question of dominance might have to be posed in terms of a deeper deterministic theory.

Professor Tisza: Unless you point out an inherent looseness in my preliminary argument (*loc. cit.*) expounding the second principle of causality, I must presume that your doubts are based on the widely accepted value judgment traceable e.g. to Boltzmann and reappearing in Einstein's critique of quantum mechanics, according to which stochastic theories are inherently second rate compared to strictly deterministic ones. This tradition, invested by great authority, brings it about that many significant contributions to statistical physics are supplemented by expressions of promise or hope, of doing better in the future, by reducing all statistical elements to rigorous deterministic dynamics. However, one of my main contentions is that this attitude originates in the lack of differentiation between the two principles of causality. Whereas the first is inconsistent with stochastic assumptions, the second is not. It seems to me that this circumstance removes a great deal of the obscurity surrounding statistical physics. As an example I will attempt to analyze the famous Einstein-Bohr controversy on the nature of quantum mechanics by breaking down the arguments into, what I believe to be their essential logical elements. According to Einstein:

E1. Nature exhibits both a space-time ordering and a causal ordering, and our theories should do likewise.

E2. The classical theories (CMP, CED, RM, RG) are the *only* ones to satisfy this requirement insofar as they operate in terms differential equations, with the exclusion of stochastic elements.

E3. Accepting *E1* and *E2* we have to expect that quantum mechanics can be recast into a form similar to that of the classical theories.

The counterarguments of Bohr, shared by Copenhagen school and by a majority of physicists, can be broken up as follows:

B1. The proposition *E1* is dismissed as an arbitrary metaphysical requirement. There is no need to expect that theory should describe "nature" or "reality", whatever these terms may mean.

B2. Identical to *E2*.

B3. Our present knowledge of quantum mechanics is inconsistent with *E3*. The essence of my own point of view is that by recognizing the bifurcation of the classical causality principle into two principles we are in a position to reject $E2 = B2$, since there are causal chains produced by stochastic laws. Granted that details have to be worked out, I believe that it will be possible to accept *E1* and *B3* and reject *B1* and *E3*. Of course, both the space-time order and the causal order have their limitations in theory and in experiment. Besides, the important morphic invariance would be impossible without these limitations.

Professor Shimony: I think in addition to the question that Professor Siegel raised about the dominance of quantum mechanics over the classical mechanics, there are some other considerations which are very troublesome in a scheme like this. Intuitively one wants quantum mechanics (being a microscopic theory) to be dominant over classical mechanics. But in all the careful formulations of quantum mechanics it is not quite that, because in the foundations of quantum mechanics itself, classical concepts enter – particularly the notion of a classical measurement; which means that the place of quantum mechanics in the scheme as it stands now is a very peculiar one. I wonder whether you think that the carrying out of this scheme to its next stage will involve considerable transformation of quantum mechanics in the classic formulation.

Professor Tisza: This is a difficult question, but not all of the difficulties stem from the obscurities of quantum mechanics. I have argued *loc. cit.* that classical mechanics consists of logically heterogeneous parts. I believe that quantum mechanics is dominant over the Hamiltonian mechanics of point masses (CMP) and also over those parts of phenomenological classical mechanics (PhCM) that are needed for the traditional theory of quantum mechanical measurements. However, quantum mechanics still has to be extended to be dominant over some subtler aspects of PhCM.

4

RUTH BARCAN MARCUS

MODAL LOGICS I: MODALITIES AND INTENSIONAL LANGUAGES*

Presented February 8, 1962

The subject of this paper is the foundations of modal logic. By foundations, we generally mean the underlying assumptions, the underpinnings.
There is a normative sense in which it has been claimed that modal logic
is without foundation. Professor Quine, in *Word and Object,* suggests that
it was conceived in sin: the sin of confusing use and mention. The original
transgressors were Russell and Whitehead. Lewis followed suit and
constructed a logic in which an operator corresponding to 'necessarily'
operates on sentences whereas 'is necessary' ought to be viewed as a
predicate of sentences. As Professor Quine reconstructs the history of the
enterprise[1], the operational use of modalities promised only one advantage: the possibility of quantifying into modal contexts. This several of
us[2] were enticed into doing. But the evils of the sentential calculus were
found out in the functional calculus, and with it – to quote again from
Word and Object – 'the varied sorrows of modality'.

I do not intend to claim that modal logic is wholly without sorrows, but
only that they are not those which Professor Quine describes. I do claim
that modal logic is worthy of defense, for it is useful in connection with
many interesting and important questions such as the analysis of causa-
tion, entailment, obligation and belief statements, to name only a few.
If we insist on equating formal logic with strongly extensional functional
calculi then Strawson[3] is correct in saying that 'the analytical equipment
(of the formal logician) is inadequate for the dissection of most ordinary
types of empirical statement.'

INTENSIONAL LANGUAGES

I will begin with the notion of an intensional language. I will make a

* This paper was announced under the more general title 'Foundation of Modal Logic'.

further distinction between those which are explicitly and implicitly intensional. Our notion of intensionality does not divide languages into mutually exclusive classes but rather orders them loosely as strongly or weakly intensional. A language is explicitly intensional to the degree to which it does not equate the identity relation with some weaker form of equivalence. We will assume that every language must have some constant objects of reference (things), ways of classifying and ordering them, ways of making statements, and ways of separating true statements from false ones. We will not go into the question as to how we come to regard some elements of experience as things, but one criterion for sorting out the elements of experience which we regard as things is that they may enter into the identity relation. In a formalized language, those symbols which name things will be those for which it is meaningful to assert that *I* holds between them, where '*I*' names the identity relation.

Ordinarily, and in the familiar constructions of formal systems, the identity relation must be held appropriate for individuals. If '*x*' and '*y*' are individual names then

(1) xIy

is a sentence, and if they are individual variables, then (1) is a sentential function. Whether a language confers thinghood on attributes, classes, propositions is not so much a matter of whether variables appropriate to them can be quantified upon (and we will return to this later), but rather whether (1) is meaningful where '*x*' and '*y*' may take as values names of attributes, classes, propositions. We note in passing that the meaningfulness of (1) with respect to attributes and classes is more frequently allowed than the meaningfulness of (1) in connection with propositions.

Returning now to the notion of explicit intensionality, if identity is appropriate to propositions, attributes, classes, as well as individuals, then any weakening of the identity relation with respect to any of these entities may be thought of as an extensionalizing of the language. By a weakening of the identity relation is meant equating it with some weaker equivalence relation.

On the level of individuals, one or perhaps two equivalence relations are customarily present: identity and indiscernibility. This does not preclude the introduction of others such as similarity or congruence, but the strongest of these is identity. Where identity is defined rather than taken as primitive, it is customary to define it in terms of indiscernibility, one form of which is

(2) $$x \text{ Ind } y =_{df} (\varphi)(\varphi x \text{ eq } \varphi y)$$

In a system of material implication (Sm), 'eq' is taken as $\equiv$. In modal systems, 'eq' may be taken as $\equiv$. In more strongly intensional systems eq may be taken as the strongest equivalence relation appropriate to such expressions as 'φx'. In separating (1) and (2) I should like to suggest the possibility that to equate (1) and (2) may already be an explicit weakening of the identity relation, and consequently an extensionalizing principle. This was first suggested to me by a paper of Ramsey.[4] Though I now regard his particular argument in support of the distinction as unconvincing, I am reluctant to reject the possibility. I suppose that at bottom my appeal is to ordinary language, since although it is obviously absurd to talk of two things being the same thing, it seems not quite so absurd to talk of two things being indiscernible from one another. In equating (1) and (2) we are saying that to be distinct is to be discernibly distinct in the sense of there being one property not common to both. Perhaps it is unnecessary to mention that if we confine things to objects with spatio-temporal histories, then it makes no sense to distinguish (1) and (2). And indeed, in my extensions of modal logic, I have chosen to define identity in terms of (2). However the possibility of such a distinction ought to be mentioned before it is obliterated. Except for the weakening of (1) by equating it with (2), extensionality principles are absent on the level of individuals.

Proceeding now to functional calculi with theory of types, an extensionality principle is of the form

(3) $$x \text{ eq } y \rightarrow xIy .$$

The arrow may be one of the implication relations present in the system or some metalinguistic conditional. 'eq' is one of the equivalence relations appropriate to x and y, but not identity. Within the system of material implication, 'x' and 'y' may be taken as symbols for classes, 'eq' as class equality (in the sense of having the same members); or 'x' and 'y' may be taken as symbols for propositions and 'eq' as the triple bar. In extended modal systems 'eq' may be taken as the quadruple bar where 'x' and 'y' are symbols for propositions. If the extended modal system has symbols for classes, 'eq' may be taken as 'having the same members' or alternatively, 'necessarily having the same members', which can be expressed within such a language. If we wish to distinguish classes from attributes in such a system (although I regard this as the perpetuation of a confusion), 'eq' may be taken as 'necessarily applies to the same thing', which is directly expressible within the system. In a language which permits epistemic contexts such as belief contexts, an even stronger equivalence relation would have to be present, than either material or strict equivalence. Taking that stronger relation as 'eq', (3) would still be an extensionalizing principle in such a strongly intensional language.

I should now like to turn to the notion of implicit extensionality, which is bound up with the kinds of substitution theorems available in a language. Confining ourselves for the sake of simplicity of exposition to a sentential calculus, one form of the substitution theorem is

$$(4) \qquad\qquad x \text{ eq}_1 y \rightarrow z \text{ eq}_2 w$$

where x, y, z, w are well-formed, w is the result of replacing one or more occurrences of x by y in z, and '$\rightarrow$' symbolizes implication or a metalinguistic conditional. In the system of material implication (Sm or QSm), (4) is provable where 'eq$_1$' and 'eq$_2$' are both taken as material equivalence for appropriate values of x, y, z, w. That is

$$(5) \qquad\qquad (x \equiv y) \supset (z \equiv w) \,.$$

Now (5) is clearly false if we are going to allow contexts involving belief, logical necessity, physical necessity and so on. We are familiar with the examples. If 'x' is taken as 'John is a featherless biped', and 'y' as 'John is a rational animal', then (5) fails. Our choice is to reject (5) as it stands, or to reject all contexts in which it fails. If the latter choice is made, the language is implicitly extensional since it cannot countenance predicates or contexts which might be permissible in a more strongly intensional language. Professor Quine's solution is the latter. All such contexts are dumped indiscriminately onto a shelf labelled 'referential opacity' or more precisely 'contexts which confer referential opacity', and are disposed of. But the contents of that shelf are of enormous interest to some of us and we would like to examine them in a systematic and formal manner. For this we need a language which is appropriately intensional.

In the modal calculus, since there are two kinds of equivalence which may hold between 'x' and 'y', (4) represents four possible substitution theorems, some of which are provable. We will return to this shortly. Similarly, if we are going to permit epistemic contexts, the modal analogue of (4) will fail in those contexts and a more appropriate one will have to supplement it.

IDENTITY AND SUBSTITUTION IN QUANTIFIED
MODAL LOGIC

In the light of the previous remarks I would like to turn specifically to the criticisms raised against extended modal systems in connection with identity and substitution. In particular, I will refer to my [5] extension of Lewis [6] $S4$ which consisted of introducing quantification in the usual manner and the addition of the axiom [7]

$$(6) \qquad \Diamond\,(\exists x)\,A \dashv\! 3\,(\exists x)\Diamond A\,.$$

I will call this system $QS4$. $QS4$ does not have an explicit axiom of extensionality, although it does have an implicit extensionalizing principle in the form of the substitution theorem. It would appear that for many uses to which modal calculi may be put, $S5$ is to be preferred.[8] In an extended $S4$, Prior[9] has shown that (6) is a theorem. My subsequent remarks, unless otherwise indicated, apply equally to $QS5$. In $QS4$ (1) is defined in terms of (2). (2), and consequently (1), admit of alternatives where 'eq' may be taken as material or strict equivalence: 'I_m' and 'I' respectively. The following are theorems of $QS4$:

$$(7) \qquad\qquad (xI_my) \equiv (xIy) \qquad\qquad \text{and}$$

$$(8) \qquad\qquad (xIy) \equiv \Box\,(xIy)$$

where '$\Box$' is the modal symbol for logical necessity. In (7) 'I_m' and 'I' are strictly equivalent. Within such a modal language, they are therefore indistinguishable by virtue of the substitution theorem. Contingent identities are disallowed by (8).

$$(9) \qquad\qquad (xIy) \cdot \Diamond \sim (xIy)$$

is a contradiction.

Professor Quine[10] finds these results offensive, for he sees (8) as 'purifying the universe.' Concrete entities are said to be banished and replaced by pallid concepts. The argument is familiar:

$$(10) \qquad\qquad \text{The evening star eq the morning star}$$

is said to express a 'true identity', yet they are not validly intersubstitutable in

(11)   It is necessary that the evening star is the evening star.

The rebuttals are also familiar. Rather than tedious repetition, I will try to restate them more persuasively. This is difficult, for I have never appreciated the force of the original argument. In restating the case, I would like to consider the following informal argument:

(12)   If p is a tautology, and p eq q, then q is a tautology

where 'eq' names some equivalence relation appropriate to p and q. In S_m if 'eq' is taken as $\equiv$ then a restricted (12) is available where $p \equiv q$ is provable.

One might say informally that with respect to any language, if (12) is said to fail, then we must be using 'tautology' in a very peculiar way, or what is taken as 'eq' is not sufficient equivalence relation appropriate to p and q. Consider the claim that

(13) aIb

is a true identity. Now if (13) is such a true identity, then a and b are the same thing. It doesn't say that a and b are two things which happen, through some accident, to be one. True, we are using two different names for that same thing, but we must be careful about use and mention. If, then, (13) is true, it must say the same thing as

(14) aIa .

But (14) is surely a tautology, and so (13) must surely be a tautology as well. This is precisely the import of my theorem (8). We would therefore expect, indeed it would be a consequence of the truth of (13), that 'a' is replaceable by 'b' in any context except those which are about the names 'a' and 'b'.

Now suppose we come upon a statement like

(15)   Scott is the author of *Waverley*

and we have a decision to make. This decision cannot be made in a formal vacuum, but must depend to a considerable extent on some informal consideration as to what it is we are trying to say in (10) and (15). If we decide that 'the evening star' and 'the morning star' are names for the same thing, and that 'Scott' and 'the author of *Waverley*' are names for the same thing, then they must be intersubstitutable in every context. In fact it often happens, in a growing, changing language, that a descriptive phrase comes to be used as a proper name – an identifying tag – and the descriptive meaning is lost or ignored. Sometimes we use certain devices such as capitalization and dropping the definite article, to indicate the change in use. 'The evening star' becomes 'Evening Star', 'the morning star' becomes 'Morning Star', and they may come to be used as names for the same thing. Singular descriptions such as 'the little corporal', 'the Prince of Denmark', 'the sage of Concord', or 'the great dissenter', are as we know often used as alternative proper names of Napoleon, Hamlet, Thoreau and Oliver Wendell Holmes. One might even devise a criterion as to when a descriptive phrase is being used as a proper name. Suppose through some astronomical cataclysm, Venus was no longer the first star of the evening. If we continued to call it alternatively 'Evening Star' or 'the evening star' then this would be a measure of the conversion of the descriptive phrase into a proper name. If, however, we would then regard (10) as false, this would indicate that 'the evening star' was not used as an alternative proper name of Venus. We might mention in passing that although the conversion of descriptions into proper names appears to be asymmetric, we do find proper names used in singular descriptions of something other than the thing named, as in the statement 'Mao Tse-tung is the Stalin of China,' where one intends to assert a similarity between the entities named.

That any language must countenance some entities as things would appear to be a precondition for language. But this is not to say that experience is given to us as a collection of things, for it would appear that there are cultural variations and accompanying linguistic variations as to what sorts of entities are so singled out. It would also appear to be a precondition of language that the singling out of an entity as a thing is accompanied by many – and perhaps an indefinite or infinite number – of unique descriptions, for otherwise how would it be singled out? But to give a thing a proper name is different from giving a unique description. For suppose we took an inventory of all the entities countenanced as things by some particular culture through its own language, with its own set of names and equatable singular descriptions, and suppose that

83

number were finite (this assumption is for the sake of simplifying the exposition). And suppose we randomized as many whole numbers as we needed for a one-to-one correspondence, and thereby tagged each thing. This identifying tag is a proper name of the thing. In taking our inventory we discovered that many of the entities countenanced as things by that language-culture complex already had proper names, although in many cases a singular description may have been used. This tag, a proper name, has no meaning. It simply tags. It is not strongly equatable with any of the singular descriptions of the thing, although singular descriptions may be equatable (in a weaker sense) with each other where

(16) $\qquad$ Desc$_1$ eq Desc$_2$

means that Desc$_1$ and Desc$_2$ describe the same thing. But here too, what we are asserting would depend on our choice of 'eq'. The principle of indiscernibility may be thought of as equating a proper name of a thing with the totality of its descriptions.

Perhaps I should mention that I am not unaware of the awful simplicity of the tagging procedure I described above. The assumption of finitude; and even if this were not assumed, then the assumption of denumerability of the class of things. Also, the assumption that all things countenanced by the language-culture complex are named or described. But my point is only to distinguish tagging from describing, proper names from descriptions. You may describe Venus as the evening star and I may describe Venus as the morning star, and we may both be surprised that as an empirical fact, the same thing is being described. But it is not an empirical fact that

(17) $\qquad$ Venus I Venus

and if 'a' is another proper name for Venus

(18) $\qquad$ Venus I a .

Nor is it extraordinary, that we often convert one of the descriptions of a thing into a proper name. Perhaps we ought to be more consistent in our use of upper-case letters, but this is a question of reforming ordinary language. It ought not to be an insurmountable problem for logicians. What I have been arguing in the past several minutes is, that to say of an identity (in the strongest sense of the word) that it is true, it must be tautologically true or analytically true. The controversial (8) of QS4 no more banishes concrete entities from the universe than (12) banishes from the universe red-blooded propositions.

Let us return now to (10) and (15). If they express a true identity, then 'Scott' ought to be anywhere intersubstitutable for 'the author of *Waverley*' and similarly for 'the morning star' and 'the evening star'. If they are not so universally intersubstitutable – that is, if our decision is that they are not simply proper names for the same thing; that they express an equivalence which is possibly false, e.g., someone else might have written *Waverley*, the star first seen in the evening might have been different from the star first seen in the morning – then they are not identities. One solution is Russell's, whose analysis provides a translation of (10) and (15) such that the truth of (10) and (15) does not commit us to the logical truth of (10) and (15), and certainly not to taking the 'eq' of (10) as identity, except on the explicit assumption of an extensionalizing axiom. Other and related solutions are in terms of membership in a non-empty unit class, or applicability of a unit attribute. But whatever the choice of a solution, it will have to be one which permits intersubstitutability, or some analogue of intersubstitutability for the members of the pairs: 'Scott' and 'the author of *Waverley*', and 'the evening star' and 'the morning star', which is short of being universal. In a language which is implicitly strongly extensional; that is where all contexts in which such substitutions fail are simply eschewed, then of course there is no harm in equating identity with weaker forms of equivalence. But why restrict ourselves in this way when, in a more intensional language, we can still make all the substitutions permissible to this weaker form of equivalence, yet admit contexts in which such a substitutivity is not permitted. To show this, I would like to turn to the instances of (4) which are provable [11] in *QS*4. I will again confine my remarks, for the purpose of exposition, to *S*4, although it is the generalizations for *QS*4 which are actually proved. An unrestricted

(19) $$x \equiv y \rightarrow z \equiv w$$

is clearly not provable whether '$\rightarrow$' is taken as material implication, strict implication or a metalinguistic conditional. It would involve us in a contradiction, if our interpreted system allowed statements such as

85

(20) $$(x \equiv y) \cdot \sim \square(x \equiv y)$$

as it must if it is not to reduce itself to the system of material implication. Indeed, the underlying assumption about equivalence which is implicit in the whole 'evening star morning star' controversy is that there are equivalences (misleadingly called 'true identities') which are contingently true. Let x and y of (19) be taken as some p and q which satisfies (20). Let z be $\square(p \equiv p)$ and w be $\square(p \equiv q)$. Then (19) is

(21) $$(p \equiv q) \rightarrow (\square(p \equiv p) \equiv \square(p \equiv q)).$$

From (20), simplification, modus ponens and $\square(p \equiv p)$, which is a theorem of $S4$, we can deduce $\square(p \equiv q)$. The latter and simplification of (20) and conjunction leads to the contradiction

(22) $$\square(p \equiv q) \cdot \sim \square(p \equiv q).$$

A restricted form of (19) is provable. It is provable if z does not contain any modal operators. And this is exactly every context allowed in Sm, without at the same time banishing modal contexts. Indeed a slightly stronger (19) is provable. It is provable if x does not fall within the scope of a modal operator in z.

Where in (4), eq_1 and eq_2 are both taken as strict equivalence, the substitution theorem

(23) $$(x \equiv y) \rightarrow (z \equiv w)$$

is provable without restriction, and also where eq_1 is taken as strict equivalence and eq_2 is taken as material equivalence as in

(24) $$(x \equiv y) \rightarrow (z \equiv w).$$

But (23) is also an extensionalizing principle, for it fails in epistemic contexts such as contexts involving 'knows that' or 'believes that'. For consider the statement

(25) When Professor Quine reviewed the paper on identity in QS4, he knew that $\vdash aI_mb \equiv aI_mb$.

and

(26) When Professor Quine reviewed the paper on identity in QS4 he knew that $\vdash aIb \equiv aI_mb$.

86

Although (25) is true, (26) is false for (7) holds in $QS4$. But rather than repeat the old mistakes by abandoning epistemic contexts to the shelf labelled 'referential opacity' after having rescued modal contexts as the most intensional permissible contexts to which such a language is appropriate, we need only conclude that (23) confines us to limits of applicability of such modal systems. If it should turn out that statements involving 'knows that' and 'believes that' permit of formal analysis, then such an analysis would have to be embedded in a language with a stronger equivalence relation than strict equivalence. Carnap's intensional isomorphism, Lewis' analytical comparability, and perhaps Anderson and Belnap's mutual entailment are attempts in that direction. But they too would be short of identity, for there are surely contexts in which substitutions allowed by such stronger equivalences, would convert a truth into a falsehood.

It is my opinion[12] that the identity relation need not be introduced for anything other than the entities we countenance as things such as individuals. Increasingly strong substitution theorems give the force of universal substitutivity without explicit axioms of extensionality. We can talk of equivalence between propositions, classes, attributes, without thereby conferring on them thinghood by equating such equivalences with the identity relation. $QS4$ has no explicit extensionality axiom. Instead we have (23), the restricted (19), and their analogues for attributes (classes). The discussion of identity and substitution in $QS4$ would be incomplete without touching on the other familiar example:

(27)   9 eq the number of planets

is said to be a true identity for which substitution fails in

(28) $\Box (9 > 7)$

for it leads to the falsehood

(29) $\Box$ (the number of planets > 7) .

Since the argument holds (27) to be contingent ($\sim \Box$ (9 eq the number of planets)), 'eq' of (27) is the appropriate analogue of material equivalence and consequently the step from (28) to (29) is not valid for the reason that the substitution would have to be made in the scope of the square. It was shown above that (19) is not an unrestricted theorem in $QS4$.
On the other hand, since in $QS4$

87

(30) $$(5 + 4) =_s 9$$

where '$=_s$' is the appropriate analogue for attributes (classes) of strict equivalence, '5 + 4' may replace '9' in (28) in accordance with (23). If, however, the square were dropped from (28) as it validly can for

(30a) $$\Box\, p \dashv 3\, p$$

is provable, then by the restricted (19), the very same substitution available to *Sm* is available here.

THE INTERPRETATION OF QUANTIFICATION

The second prominent area of criticism of quantified modal logic involves interpretation of the operations of quantification when combined with modalities. It appears to me that at least some of the problems stem from an absence of an adequate, unequivocal, colloquial translation of the operations of quantification. It is often not quantification but our choice of reading and implicit interpretive consequences of such a reading which leads to difficulties. Such difficulties are not confined to modal systems. The most common reading of existential quantification is

(31)   There is (exists) at least one (some) thing (person) which (who)...

Strawson[13], for example, does not even admit of significant alternatives, for he says of (31): '...we might think it strange that the whole of modern formal logic after it leaves the propositional logic and before it crosses the boundary into the analysis of mathematical concepts, should be confined to the elaboration of sets of rules giving the logical interrelations of formulae which, however complex, begin with these few rather strained and awkward phrases.' Indeed, taking (31) at face value, Strawson gets into a muddle about tense ((31) is in the present tense), and the ambiguities of the word 'exist'. What we would like to have and do not have, is a direct, unequivocal colloquial reading of

(32) $$(\exists x)\ \varphi x$$

which gives us the force of either of the following:

(33)   Some substitution instance of φx is true

or

There is at least one value of x for which φx is true.

I am not suggesting that (33) provides translations of (32), but only that what is wanted is a translation with the force of (32).

As seen from (33), quantification has primarily to do with truth and falsity, and open sentences. Reading in accordance with (31) may entangle us unnecessarily in ontological perplexities. For if quantification has to do with things and if variables for attributes or classes can be quantified upon, then in accordance with (31) they are things. If we still want to distinguish the identifying from the classifying function of language, then we are involved in a classification of different kinds of things and the accompanying platonic involvements. The solution is not to banish quantification on variables other than individual variables, but only not to be taken in by (31). We do in fact have some colloquial counterparts of (33). The non-temporal 'sometimes' or 'in some cases' or 'in at least one case', which have greater ontological neutrality than (31).

Some of the arguments involving modalities and quantification are closely connected with questions of substitution and identity. At the risk of boredom I will go through one again. In $QS4$ the following definitions are introduced [14]:

(34) $$(\varphi =_m \Psi) =_{df} (x)(\varphi x \equiv \Psi x)$$

(35) $$(\varphi =_s \Psi) =_{df} \square(\varphi =_m \Psi)$$

Since the equality in (10) is contingent, (10) may be written as

(36) $\qquad$ (the evening star $=_m$ the morning star) .

It is also the case that

(37) $\qquad$ $\Diamond \sim$ (the evening star $=_m$ the morning star) .

One way of writing (11) is as

(38) $\qquad$ $\square$ (the evening star $=_m$ the evening star) .

By existential generalization on (38), it follows that

(39) $\qquad$ $(\exists\,\varphi)\,\square\,(\varphi =_m$ the evening star) .

In the words of (31), (39) becomes

(40) $\quad$ There is a thing such that it is necessary that it is equal to the evening star.

89

The stubborn unlaid ghost rises again. Which thing, the evening star which by (36) is equal to the morning star? But such a substitution would lead to the falsehood

(41) $\square$ (the evening star $=_m$ the morning star) .

The argument may be repeated for (27) through (29).

In $QS4$ the solution is clear. For since (37) holds, and since in (39) 'φ' occurs within the scope of a square, then we cannot go from (39) to (41). On the other hand the step from (38) to (39) (existential instantiation) is entirely valid. For surely there is a value of φ for which

$$\square (\varphi = \text{the evening star})$$

is true. In particular, the case where 'φ' is replaced by 'the evening star'.

There is also the specific problem of interpreting quantification in (6), which is a postulate of $QS4$. Read in accordance with (31) as

(42) If it is logically possible that there is something which φ's, then there is something such that it is logically possible that it φ's ,

it is admittedly odd. The antecedent seems to be about what is logically possible and the consequent about what there is. How can one go from possibility to existence? Read in accordance with (33) we have the clumsy but not so paradoxical

(43) If it is logically possible that φx for some value of x, then there is some value of x such that it is logically possible that φx .

Although the emphasis has now been shifted from things to statements, and the ontological consequences of (42) are absent, it is still indirect and awkward. It would appear that questions such as the acceptability or non-acceptability of (6) are best solved in terms of some semantical construction. This will be returned to in conclusion, but first some minor matters.

A defense of modal logic would be incomplete without touching on criticisms of modalities which stem from confusion about what is or isn't provable in such systems. One example is that of Rosenbloom [15] who seized on the fact that a strong deduction theorem is not available in

$QS4$, as a reason for discarding strict implication as in any way relevant to the deducibility relation. He failed to note [16] that a weaker and perhaps more appropriate deduction theorem is available. Indeed, Anderson and Belnap [17], in their attempt to formalize entailment without modalities, reject the strong form of the deduction theorem as 'counter-intuitive for entailment'.

Another example occurs in *Word and Object* [18] which can be summarized as follows:

(44) Modalities yield talk of a difference between necessary and contingent attributes.

(45) Mathematicians may be said to be necessarily rational and not necessarily two-legged.

(46) Cyclists are necessarily two-legged and not necessarily rational.

(47) a is a mathematician and a cyclist.

(48) Is this concrete individual necessarily rational or contingently two-legged or vice versa?

(49) 'Talking referentially of the object with no special bias toward a background grouping of mathematicians as against cyclists... there is no semblance of sense in rating some of his attributes as necessary and others as contingent.'

Professor Quine says that (44) through (47) are supposed to 'evoke the appropriate sense of bewilderment' and they surely do. For I know of no interpreted modal system which countenances necessary attributes in the manner suggested. Translating (45) through (47) we have

(50) $(x)(Mx \dashv_3 Rx) \equiv (x) \square (Mx \supset Rx) \equiv (x) \sim \lozenge (Mx \cdot \sim Rx)$

which is conjoined in (45) with

(51) $(x) \sim \square (Mx \supset Tx) \equiv (x) \lozenge \sim (Mx \supset Tx) \equiv (x) \lozenge (Mx \cdot \sim Tx)$.

Also

(52) $(x)(Cx \dashv_3 Tx) \equiv (x) \square (Cx \supset Tx) \equiv (x) \sim \lozenge (Cx \cdot \sim Tx)$

which is conjoined in (46) with

(53) $(x) \sim \square (Cx \supset Rx) \equiv (x) \diamondsuit \sim (Cx \supset Rx) \equiv (x) \diamondsuit (Cx \cdot \sim Rx)$.

And in (48)

(54)
$$Ma \cdot Ca$$

Among the conclusions we can draw from (49) through (53) are

$\square (Ma \supset Ra)$, $\sim \diamondsuit (Ma \cdot \sim Ra)$, $\diamondsuit (Ma \cdot \sim Ta)$, $\sim \square (Ma \supset Ta)$,
$\square (Ca \supset Ta)$, $\sim \diamondsuit (Ca \cdot \sim Ta)$, $\diamondsuit (Ca \cdot \sim Ra)$, $\sim \square (Ca \cdot \sim Ra)$,
$Ta, Ra, Ta \cdot Ra$

But nothing to answer question (48), or to make any sense of (49). It would appear that Professor Quine is assuming

(55)
$$(p \dashv 3 \, q) \dashv 3 (p \dashv 3 \, \square \, q)$$

is provable in $QS4$, but it is not, except where $p \equiv \square \, r$ for some r. Keeping in mind that we are dealing with logical modalities, none of the attributes in (50) through (54) taken separately, or conjoined, are necessary. It is not that sort of attribute which modal logic, even derivatively, countenances as being necessary. A word is appropriate here about the derivative sense in which we can speak of logically necessary and contingent attributes.

In $QS4$ abstracts are introduced such that to every function there corresponds an abstract, e.g.

(56) $x \varepsilon \hat{y} A =_{df} B$, where B is the result of substituting every free occurrence of y in A by x.

If r is some abstract then we can define

(57) $x \varepsilon \, \boxdot \, r =_{df} \square (x \varepsilon r)$, $\vdash \boxdot \, r =_{df} (x)(x \varepsilon \, \boxdot \, r)$

and

(58) $x \varepsilon \diamondsuit r =_{df} \diamondsuit (x \varepsilon r)$, $\vdash \diamondsuit \, r =_{df} (x)(x \varepsilon \diamondsuit r)$

It is clear that among the abstracts to which $\vdash \boxdot$ may validly be affixed, will be those corresponding to tautological functions, e.g., $\hat{y}(yIy)$, $\hat{y}(\varphi x \lor \sim \varphi x)$, etc. It would be appropriate to call these necessary attributes, and the symbol '$\boxdot$' is a derivative way of applying modalities to attributes.

92

Similarly, all of the attributes of (50) through (54) could in the sense of (58) be called contingent, where '$\Diamond$' is the derivative modality for contingency of attributes. However, if (50) is true, then the attribute of being either a mathematician or not rational could appropriately be called necessary, for

(59) $\qquad\qquad (x) \, \Box \, (x\varepsilon\hat{y}(Mx \lor \sim Rx)) \, .$

SEMANTIC CONSTRUCTIONS

I would like in conclusion to suggest that the polemics of modal logic are perhaps best carried out in terms of some explicit semantical construction. As we have seen in connection with (6) it is awkward at best and at worst has the character of a quibble, not to do so.

Let us reappraise (6) in terms of such a construction.[19] Consider for example a language (L), with truth functional connectives, a modal operator ($\Diamond$), a finite number of individual constants, an infinite number of individual variables, one two-place predicate (R), quantification and the usual criteria for being well-formed. A domain (D) of individuals is then considered which are named by the constants of L. A model of L is defined as a class of ordered couples (possibly empty) of D. The members of a model are exactly those pairs between which R holds. To say therefore that the atomic sentence $R(a_1a_2)$ of L holds or is true in M, is to say that the ordered couple (b_1, b_2) is a member of M, where a_1 and a_2 are the names in L of b_1 and b_2. If a sentence A of L is of the form $\sim B$, A is true in M if and only if B is not true in M. If A is of the form $B_1 \cdot B_2$ then A is true in M if and only if both B_1 and B_2 are true in M. If A is of the form $(\exists x) \, B$, then A is true in M if and only if at least one substitution instance of B is true (holds) in M. If A is $\Diamond \, B$ then A is true in M if and only if B is true in some model M_1.

We see that a true sentence of L is defined relative to a model and a domain of individuals. A logically true sentence is one which would be true in every model. We are now in a position to give a rough proof of (6). Suppose (6) is false in some M. Then

$$\sim (\Diamond \, (\exists x) \, \varphi x \cdot \sim (\exists x) \, \Diamond \, \varphi x)$$

is false in M. Therefore

$$\Diamond (\Diamond (\exists x)\, \varphi x \cdot \sim (\exists x)\, \Diamond \, \varphi x)$$

is true in M. So

$$\Diamond (\exists x)\phi x \cdot \sim (\exists x)\, \Diamond \, \phi x$$

is true in some M_1. Therefore

(60) $$\Diamond (\exists x)\, \varphi x$$

and

(61) $$\sim (\exists x)\, \Diamond \, \varphi x$$

are true in M_1. Consequently, from (60)

(62) $$(\exists x)\, \varphi x$$

is true in some model M_2. Therefore there is a member of D (b) such that

(63) $$\varphi b$$

is true in M_2. But from (61)

$$(\exists x)\, \Diamond \, \varphi x$$

is not true in M_1. Consequently there is no member of D such that

(64) $$\Diamond \, \varphi b$$

is true in M_1. So there is no model M_2 such that φb is true in M_2. But this result contradicts (63). Consequently, in such a construction, (6) must be true in every model.

If this is the sort of construction one has in mind then we are persuaded of the plausibility of (6). Indeed, going back to (43), it can be seen that this was the sort of construction which was being assumed. If (6) is to be regarded as offensive in a way other (and here I am borrowing an image from Professor White) than the manner in which we regard eating peas with a knife as offensive, it must be in terms of some semantic construction which ought to be made explicit.[20]

We see, that though the rough outline above corresponds to the Leibnizian distinction between true in a possible world and true in all possible worlds, it is also to be noted that there are no specifically intensional objects. No new entity is spawned in a possible world that isn't already in the domain in terms of which the class of models is defined. In such a model modal operators have to do with truth relative to the model, not with things. On this interpretation,[21] Professor Quine's 'flight from intension' may have been exhilarating, but unnecessary.[22]

Department of Philosophy, Roosevelt University, Chicago, Illinois

NOTES

1. W. V. Quine, *Word and Object*, 1960, pp. 195–196.
2. a. F. B. Fitch, *Symbolic Logic*, New York, 1952.
   b. R. Carnap, Modalities and quantification, *Journal of Symbolic Logic* 11 (1946) 33–64.
   c. R. C. Barcan (Marcus), A functional calculus of first order based on strict implication, *Journal of Symbolic Logic* 11 (1946) 1–16.
   d. R. C. Barcan (Marcus), The identity of individuals in a strict functional calculus of first order, *Journal of Symbolic Logic* 12 (1947) 12–15.
3. P. F. Strawson, *Introduction to Logical Theory*, London, 1952, p. 216.
4. F. P. Ramsey, *The Foundations of Mathematics*, London and New York, 1931, pp. 30–32.
5. *Op. cit*, notes 2c, 2d.
6. C. I. Lewis and C. H. Langford, *Symbolic Logic*, New York, 1932.
7. See A. N. Prior, *Time and Modality*, Oxford, 1932, for an extended discussion of this axiom.
8. S5 results from adding to S4.

$$p \dashv 3 \ \square \ \Diamond \ p.$$

9. A. N. Prior, Modality and Quantification in S5, *Journal of Symbolic Logic* 21 (1956).
10. W. Van Orman Quine, *From a Logical Point of View*, Cambridge, 1953, pp. 152–154.
11. *Op. cit.*, note 2c. Theorem XIX* corresponds to (23). The restricted (19), given the conditions of the restriction, although not actually proved, is clearly provable in the same manner as XIX*.
12. See R. Barcan Marcus, Extensionality, *Mind* 69, 55–62 which overlaps to some extent the present paper.
13. *Op. cit.*, note 3.
14. *Op. cit.*, note 2c. Abstracts are introduced and attributes (classes) may be equated with abstracts. Among the obvious features of such a calculus of attributes (classes), is the presence of equivalent, non-identical, empty attributes (classes). If the null attribute (class) is defined in terms of identity, then it will be intersubstitutible with any abstract on a contradictory function.
15. P. Rosenbloom, *The Elements of Mathematical Logic*, New York, 1950, p. 60.
16. R. Barcan Marcus, Strict implication, deducibility, and the deduction theorem, *Journal of Symbolic Logic* 18 (1953) 234–236.

17. A. R. Anderson and N. D. Belnap, *The Pure Calculus of Entailment*, (pre-print).
18. *Op. cit.*, note 1, pp. 199–200.
19. The construction here outlined corresponds to that of R. Carnap, *Meaning and Necessity*, Chicago, 1947. The statement of the construction is in accordance with the method of J. C. C. McKinsey. See also, J. C. C. McKinsey, On the syntactical construction of systems of modal logic, *Journal of Symbolic Logic* 10 (1946) 88–94; A new definition of truth, *Synthese* 7 (1948–1949) 428–433.
20. A criticism of the construction here outlined is the assumption of the countability of members of D. McKinsey points this out in the one chapter I have seen (Chapter I, Vol. II), of a projected (unpublished) two volume study of modal logic, and indicates that his construction will not assume the countability of members of D. Whereas Carnap's construction leads to a system at least as strong as S5, McKinsey's (he claims) will be at least as strong as S4 (without (6) I would assume). I've not seen, nor been able to locate any other parts of this study in which the details are worked out along with completeness proofs for some of the Lewis systems. See also, J. Myhill, *Logique et analyse*, 1958, pp. 74–83; and S. Kripke, *Journal of Symbolic Logic* 24 (1959) 323–324 (abstract).
22. If one wishes to talk about possible things then of course such a construction is inadequate.
22. This paper was written while the author was under N.S.F. Grant 24335.

COMMENTS

WILLARD VAN ORMAN QUINE

Professor Marcus struck the right note when she represented me as suggesting that modern modal logic was conceived in sin: the sin of confusing use and mention. She rightly did not represent me as holding that modal logic *requires* confusion of use and mention. My point was a historical one, having to do with Russell's confusion of 'if-then' with 'implies'.

Lewis founded modern modal logic, but Russell provoked him to it. For whereas there is much to be said for the material conditional as a version of 'if-then', there is nothing to be said for it as a version of 'implies'; and Russell called it implication, thus apparently leaving no place open for genuine deductive connections between sentences. Lewis moved to save the connections. But his way was not, as one could have wished, to sort out Russell's confusion of 'implies' with 'if-then'. Instead, preserving that confusion, he propounded a strict conditional and called *it* implication.

It is logically possible to like modal logic without confusing use and mention. You could like it because, apparently at least, you can quantify into a modal context by a quantifier outside the modal context, whereas you obviously cannot coherently quantify into a mentioned sentence from outside the mention of it. Still, man is a sensemaking animal, and as such he derives little comfort from quantifying into modal contexts that he does not think he understands. On this score, confusion of use and mention seems to have more than genetic significance for modal logic. It seems to be also a sustaining force, engendering an illusion of understanding.

I am speaking empirically. There was a period twenty-five years ago when I kept being drawn into arguments with C. I. Lewis and E. V. Huntington over interpretation of modal logic; and in those arguments I found it necessary to harp continually on the theme of use versus mention. And now points that Professor Marcus has urged this evening, in favor of modal logic, force me back to that same theme again.

Thus consider her 'informal argument:

(12) If p is a tautology, and p eq q, then q is a tautology'.

Her adoption of the letters 'p' and 'q', rather than say 'S₁' and 'S₂', suggests that she intends them to occupy sentence positions. Also her 'eq' is perhaps intended as a sentence *connective*, despite her saying that it names some equivalence relation; for she says that it could be taken as '≡'. On the other hand her clauses 'p is a tautology' and 'q is a tautology' do not show 'p' and 'q' in sentence position. These clauses show 'p' and 'q' in name positions, as if they were replaceable not by sentences but by names of sentences.

Or try the opposite interpretation. Suppose that Professor Marcus contrary to custom, is using 'p' and 'q' as variables whose values are sentences, and whose proper substitutes are therefore names of sentences. Then 'eq' is indeed to be seen as naming some equivalence *relation*, just as she says; and the mention of '≡' must be overlooked as an inadvertency. On this interpretation, (12) is unexceptionable. But on this interpretation (12) is no part of modal logic; it is ordinary non-modal metalogic. For on this interpretation 'eq' is not a non-truth-functional sentence connective at all, but an ordinary non-truth-functional two-place sentence predicate, like 'implies'. I have no objection to these. In my logical writings early and late I have used them constantly.

Twenty-five years ago, in arguing much the same matter with Lewis and Huntington at vastly greater length, I was forced to recognize my inability to make people aware of confusing use and mention. Nor have the passing years brought me the ability; they have only vindicated my despair of it. By now perhaps I should have concluded that I must be the confused one, were it not for people who do turn out to see the distinction my way. I have said that modal logic does not require confusion of use and mention. But there is no denying that confusion of use and mention engenders an irresistible case for modal logic, as witness (12).

I should not leave (12) without touching upon a third interpretation. Perhaps 'p' and 'q' are to be seen as propositional variables, whose values are propositions (or meanings of sentences) and whose appropriate substitutions are therefore names of propositions, hence names of meanings of sentences. Then again (12) is in order, if we countenance these subtle entities. But, on this interpretation, 'eq' comes to name a relation between propositions; again it is no connective of sentences. To suppose it were would be to confuse meaning with reference, and thus to view sentences as names of their meanings.

Let me move now to Professor Marcus's discussion of her (13) and (14), viz. 'aIb' and 'aIa'. Suppose that aIb. Then, she argues, anything true of a is true of b. I agree. But, she says, 'aIa' is a tautology. Again I agree, not quarreling over the term. So, she concludes, 'aIb' must be a tautology too. Why? The reasoning is as follows. We are trying to prove this about b: not just that aIb, but that tautologously aIb. Now this thing that we are trying to prove about b, viz., that tautologously aIb, is true of a; so, since b is a, it is true of b.

Again our troubles condense about the distinction between use and mention. If we take 'tautologously' as a modal operator attachable directly to sentences, then the argument is all right, but pointless so long as the merits of modal logic are under debate. If on the other hand we accept only 'tautologous', as a predicate attributable to sentences and therefore attachable to quotations of sentences, then the argument breaks down. For, the property that was to be proved about b – viz., that tautologously aIb – has to be seen now as a quotation-breaking pseudo-property on which the substitutivity of identity has no bearing. What I mean by a quotation-breaking pseudo-property will be evident if we switch for a moment to the truth ''Cicero' has three syllables'. Obviously we cannot infer that 'Tully' has three syllables, even though Tully is Cicero. And from 'aIa' is tautologous there is no more reason to infer that ''aIb' is tautologous', even granted that b is a.

Professor Marcus's reflections on identity led her to conclude that identity, substitutivity, and extensionality are things that come in grades. I have just now objected to some of the reasoning. I also do not accept the conclusion. My position is that we can settle objectively and absolutely what predicate of a theory to count as the identity predicate, if any, once we have settled what notations to count as quantifiers, variables, and the truth functions. Until we have found how to handle quantification in a given theory, of course we have no way even of telling what expressions of the theory to count as predicates and what signs to count as their subject variables; and, not being able to spot predicates, we cannot spot the identity predicate. But show me the quantifiers and the variables and the truth functions, and I can show you when to read an open sentence 'ϕxy' as 'x $=$ y'. The requirements are strong reflexivity and substitutivity, thus:

$$(x)\phi xx, \quad (x)(y)(\phi xy \cdot \ldots x \ldots \cdot \supset \cdot \ldots y \ldots).$$

If these requirements are met, then, as is well known, 'ϕxy' meets all the formal requirements of 'x $=$ y'; and otherwise not.

The requirements fix identity uniquely. That is, if 'ϕ' and 'ψ' both meet the requirements of strong reflexivity and substitutivity, then they are coextensive. Let me quickly prove this. By substitutivity of 'ϕ',

$$(x)(y)(\phi xy \cdot \psi xx \cdot \supset \psi xy).$$

But, by reflexivity of ψ, we can drop the 'ψxx'. So 'ψ' holds wherever 'ϕ' does. By the same argument with 'ϕ' and 'ψ' interchanged, 'ϕ' holds wherever 'ψ' does.

There are a couple of tangents that I would just mention and not use. One is that there is no assurance, given a theory with recognized notations for quantification and the truth functions, that there is an identity predicate in it. It can happen that no open sentence in 'x' and 'y', however complex, is strongly reflexive and substitutive. But this is unusual.

The other is that if an open sentence in 'x' and 'y' does meet these two requirements, we may still find it to be broader than true identity when we interpret it in the light of prior interpretations of the primitive predicates of the theory. But this sort of discrepancy is always traceable to some gratuitous distinctions in those prior interpretations of the primitive predicates. The effect of our general rule for singling out an identity predicate is a mild kind of identification of indiscernibles.[1]

Tangents aside, my point is that we have an objective and unequivocal criterion whereby to spot the identity predicate of a given theory, if such there be. The criterion is independent of what the author of the theory may do with '$=$' or 'I' or the word 'identity'. What it does depend on is recognition of the notations of quantification and the truth functions. The absoluteness of this criterion is important, as giving a fixed point of reference in the comparison of theories. Questions of universe, and individuation, take on a modicum of inter-systematic significance that they would otherwise lack.

In particular the criterion makes no doubt of Professor Marcus's law for modal logic:

$$(x)(y)(x = y \cdot \supset \cdot \text{necessarily } x = y).$$

It follows from 'necessarily $x = x$' by substitutivity.

Notice that my substitutivity condition was absolute. There was no question what special positions to exempt from substitutivity, and no question what special names or descriptions to exempt in special positions. Hence there was no scope for gradations of identity or substitutivity. What enabled me to cut so clean was that I talked in terms not of names or descriptions but of 'x' and 'y': variables of quantification. The great philosophical value of the eliminability of singular terms other than variables is that we can sometimes thus spare ourselves false leads and lost motion.

In her own continuing discussion, Professor Marcus developed a contrast between proper names and descriptions. Her purpose was, I gather, to shed further light on supposed grades or alternatives in the matter of identity and substitutivity. I have urged just now that we can cut through all this by focusing on the bindable variable. And I am glad, for I think I see trouble anyway in the contrast between proper names and descriptions as Professor Marcus draws it. Her paradigm of the assigning of proper names is tagging. We may tag the planet Venus, some fine evening, with the proper name 'Hesperus'. We may tag the same planet again, some day before sunrise, with the proper name 'Phosphorus'. When at last we discover that we have tagged the same planet twice, our discovery is empirical. And not because the proper names were descriptions.

In any event, this is by the way. The contrast between description and name needs not concern us if we take rather the variables of quantification as our ultimate singular terms. Already for the second time we note the philosophical value of the eliminability of singular terms other than variables: again it spares us false leads and lost motion.

Let us look then to Professor Marcus's next move. Alarmingly, her next move was to challenge quantification itself, or my object-oriented interpretation of it. Here she talks of values of variables in a sense that I must sharply separate from my own. For me the values e.g. of number variables in algebra are not the numerals that you can substitute, but the

101

numbers that you talk about. For Professor Marcus, the values are the expressions you can substitute. I think my usage has the better history, but hers has a history too. Ryle objected somewhere to my dictum that to be is to be the value of a variable, arguing that the values of variables are expressions and hence that my dictum repudiates all things except expressions. Clearly, then, we have to distinguish between values of variables in the *real* sense and values of variables in the *Ryle* sense. To confuse these is, again, to confuse use and mention. Professor Marcus is not, so far as I observe, confusing them. She simply speaks of values of variables in the Ryle sense. But to forestall confusion I should like to say 'substitutes for variables' rather than 'values of variables' in this sense, thus reserving 'values of variables' for values of variables in the real sense.

Thus paraphrased, Professor Marcus's proposed reinterpretation of existential quantification is this: the quantification is to be true if and only if the open sentence after the quantifier is true for some substitute for the variable of quantification. Now this is, I grant, an intelligible reinterpretation, and one that does not require objects, in any sense, as values, in the real sense, of the variables of quantification. Note only that it deviates from the ordinary interpretation of quantification in ways that can matter. For one thing, there is a question of unspecifiable objects. Thus take the real numbers. On the classical theory, at any rate, they are indenumerable, whereas the expressions, simple and complex, available to us in any given language are denumerable. There are therefore, among the real numbers, infinitely many none of which can be separately specified by any expression, simple or complex. Consequently an existential quantification can come out true when construed in the ordinary sense, thanks to the existence of appropriate real numbers, and yet be false when construed in Professor Marcus's sense, if by chance those appropriate real numbers all happen to be severally unspecifiable.

But the fact remains that quantification can indeed be thus reinterpreted, if not altogether *salva veritate*, so as to dissociate it from objective reference and real values of variables. Why should this be seen as desirable? As an answer, perhaps, to the charge that quantified modal logic can tolerate only intensions and not classes or individuals as values of its variables? But it is a puzzling answer. For, it abstracts from reference altogether. Quantification ordinarily so-called is purely and simply the logical idiom of objective reference. When we reconstrue it in terms of substituted expressions rather than real values, we waive reference. We preserve distinctions between true and false, as in truth-function logic itself, but we cease to depict the referential dimension. Now anyone who is willing to abstract thus from questions of universe of discourse cannot have cared much whether there were classes and individuals or only intensions in the universe of discourse. But then why the contortions? In short, if reference matters, we cannot afford to waive it as a category; and if it does not, we do not need to.

As a matter of fact, the worrisome charge that quantified modal logic can tolerate only intensions and not classes or individuals was a mistake to begin with. It goes back to 1943; my 'Notes on existence and necessity'[2] and Church's review of it.[3] To illustrate my misgivings over quantifying into modal contexts I used, in that article, the example of 9 and the number of the planets. They are the same thing, yet 9 necessarily exceeds 7 whereas the number of the planets only contingently exceeds 7. So, I argued, necessarily exceeding 7 is no trait of the neutral thing itself, the number, which is the number of the planets as well as 9. And so it is nonsense to say neutrally that there is *something*, x, that necessarily exceeds 7. Church countered that my argument worked only for things like numbers, bodies, classes, that we could specify in contingently coincident ways: thus 9 is what succeeds 8, and is what numbers the planets, and these two specifications only contingently coincide. If we limit our objects to intensions, Church urged, this will not happen.

Now on this latter point Church was wrong. I have been slow to see it, but the proof is simple. Anything x, even an intension, is specifiable in contingently coincident ways if specifiable at all. For, suppose x is determined uniquely by the condition 'ϕx'. Then it is also determined uniquely by the conjunctive condition '$p \cdot \phi x$' where 'p' is any truth, however irrelevant. Take 'p' as an arbitrary truth not implied by 'ϕx', and these two specifications of x are seen to be contingently coincident: 'ϕx' and '$p \cdot \phi x$'.

Contrary to what Church thought, therefore, my 1943 strictures were cogent against quantification over any sorts of objects if cogent at all; nothing is gained by limiting the universe to intensions. The only course open to the champion of quantified modal logic is to meet my strictures head on: to argue in the case of 9 and the number of the planets that this number is, of itself and independently of mode of specification, something that necessarily, not contingently, exceeds 7. This means adopting a frankly inequalitarian attitude toward the various ways of specifying the number. One of the determining traits, the succeeding of 8, is counted as a necessary trait of the number. So are any traits that follow from that one, notably the exceeding of 7. Other uniquely determining traits of the number, notably its numbering the planets, are discounted as contingent traits of the number and held not to belie the fact that the number does still necessarily exceed 7.

This is how essentialism comes in: the invidious distinction between some traits of an object as essential to *it* (by whatever name) and other traits of it as accidental. I do not say that such essentialism, however uncongenial to me, should be uncongenial to the champion of quantified modal logic. On the contrary, it should be every bit as congenial as quantified modal logic itself.[4]

Harvard University, Cambridge, Massachusetts

NOTES

1. Quine, *Word and Object*, New-York, 1960, p. 230.
2. *Journal of Philosophy* **40** (1943) 113–127.
3. A. Church, *Journal of Symbolic Logic* **8** (1943) 45–47.
4. For more in the vein of these last few paragraphs see my *From a Logical Point of View*, Revised Edition, Cambridge, Mass., 1961, pp. 148–157.

DISCUSSION

RUTH BARCAN MARCUS, WILLARD VAN ORMAN QUINE,

SAUL KRIPKE, JOHN MCCARTHY,

DAGFINN FOLLESDAL

Prof. Marcus: We seem still at the impasse I thought to resolve at this time. The argument concerning (12) was informal, and parallels as I suggested, questions raised in connection with the 'paradox' of analysis. One would expect that if a statement were analytic, and it bore a strong equivalence relation to a second statement, the latter would be analytic as well. Since (12) cannot be represented in S_m without restriction, the argument reveals material equivalence to be insufficient and weak. An adequate representation of (12) requires a modal framework.

The question I have about essentialism is this: Suppose these modal systems are extended in the manner of *Principia* to higher orders. Then

$$\Box ((5 + 4) = 9)$$

is a theorem ('=' here may be taken as either '$=_s$' or '$=_m$' of the present paper, since the reiterated squares telescope), whereas

$$\Box ((5 + 4) = \text{the number of planets})$$

is not. Our interpretation of these results commits us only to the conclusion that the equivalence relation which holds between $5 + 4$ and 9 is stronger than the one which holds between $5 + 4$ and the number of planets. More specifically, the stronger one is the class or attribute analogue of $\equiv$. No mysterious property is being conferred on either 9 or the number of planets which it doesn't already have in the extensional $((5 + 4) =_m \text{the number of planets})$.

Prof. Quine: May I ask if Kripke has an answer to this?... Or I'll answer, or try to.

Mr. Kripke: As I understand Professor Quine's essentialism, it isn't what's involved in either of these two things you wrote on the board, that causes trouble. It is in inferring that there exists an x, which necessarily = 5 + 4 (from the first of the two). (to Quine:) Isn't that what's at issue?

Prof. Quine: Yes.

Mr. Kripke: So this attributes necessarily equalling 5 + 4 to an object.

Prof. Marcus: But that depends on the suggested interpretation of quantification. We prefer a reading that is not in accordance with things, unless, as in the first order language there are other reasons for reading in accordance with things.

Prof. Quine: That's true.

Prof. Marcus: So the question of essentialism arises only on your reading of quantification. For you, the notion of reference is univocal, absolute, and bound up with the expressions, of whatever level, on which quantification is allowed. What I am suggesting is a point of view which is not new to the history of philosophy and logic. That all terms may refer to objects, but that not all objects are things where a thing is at least that about which it is appropriate to assert the identity relation. We note a certain historical consistency here, as for example, the reluctance to allow identity as a relation proper to propositions. If one wishes, one could say that object-reference (in terms of quantification) is a wider notion than thing-reference, the latter being also bound up with identity and perhaps with other restrictions as well such as spatio-temporal location. If one wishes to use the word 'refer' exclusively for thing-reference, then we would distinguish those names which refer, from those which name other sorts of objects. Considered in terms of the semantical construction proposed at the end of the paper, identity is a relation which holds between individuals; and their names have thing-reference. To say of a thing a that it necessarily has a property φ $(\Box \, (\varphi a))$, is to say that $\varphi \, a$ is true in every model. Self-identity would be such a property.

Prof. Quine: Speaking of the objects or the referential end of things in terms of identity, rather than quantification, is agreeable to me in the sense that for me these are inter-definable anyway. But what's appropriately regarded as the identity matrix, or open sentence, in the theory is for me determined certainly by consideration of quantification. Quantification is a little bit broader, a little bit more generally applicable to the theory because you don't always have anything that would fulfil this identity requirement. As to where essentialism comes in: what I have in mind is an interpretation of this quantification where you have an x here (in $\Box$

$((5 + 4) = x))$. Now I appreciate that from the point of view of modal logic, and of things that have been done in modal logic in Professor Marcus' pioneer system, this would be regarded as true rather than false:

$$\Box ((5 + 4) = 9)$$

This is my point, in spite of the fact that if you think of this ($\Box (5 + 4) =$ the number of planets) as what it is generalized from, it ought to be false.

Prof. Marcus:

$$\Box ((5 + 4) = \text{the number of planets})$$

would be false. But this does not preclude the truth of

$$(\exists x) \Box (5 + 4) = x) =$$

anymore than the falsehood

$$12 = \text{the number of Christ's disciples}$$

precludes the truth of

$$(\exists x) (12 = x) .$$

(We would, of course, take '=' as '$=_m$' here.)

Prof. Quine: That's if we use quantification in the ordinary ontological way and that's why I say we put a premium on the *nine* as over against the *number of planets*; we say this term is what is going to be *maßgebend* for the truth value of this sentence in spite of the fact that we get the opposite whenever we consider the other term. This is the sort of specification of the number that counts:

$$5 + 4 = 9$$

This is not: $5 + 4 = $ number of planets.

I grant further that essentialism does not come in if we interpret quantification in your new way. By quantification I mean, quantification in the ordinary sense rather than a new interpretation that might fit most if not all of the formal laws that the old quantification fits. I say 'if not all', because I think of the example of real numbers again. If on the other hand we do not have quantification in the old sense then I have nothing to suggest at this point about the ontological implications or difficulties of modal logic. The question of ontology wouldn't arise if there were no quantification of the ordinary sort. Furthermore, essentialism certainly wouldn't be to the point, for the essentialism I'm talking about is essentialism in the sense that talks about objects, certain objects; that an object has certain of these attributes essentially, certain others only accidentally. And no such question of essentialism arises if we are only talking of the terms and not the objects that they allegedly refer to. Now Professor Marcus also suggested that possibly the interpretation could be made something of a hybrid between the two – between quantification thought of as a formal matter, and just talking in a manner whose truth conditions are set up in terms of the expression substituted rather than in terms of the objects talked about; and that there are other cases where we can still give quantification the same old force. Now that may well be: we might find that in the ordinary sense of quantification I've been talking about there is quantification into non-modal contexts and no quantification but only this sort of quasi-quantification into the modal ones. And this conceivably might be as good a way of handling such modal matters as any.

Prof. Marcus: It is not merely a way of coping with perplexities associated with intensional contexts. I think of it as a better way of handling quantification.

You've raised a problem which has to do with the real numbers. Perhaps the Cantorian assumption is one we can abandon. We need not be particularly concerned with it here.

Prof. Quine: It's one thing I would certainly be glad to avoid, if we can get all of the classical mathematics that we do want.

Mr. Kripke: This is what I thought the issue conceivably might be, and hence I'll raise it explicitly in this form: Suppose this system contains names, and suppose the variables are supposed to range over numbers, and using "9" as the name of the number of planets, and the usual stock of numerals, "0", "1", "2", , , and in addition various other primitive terms for numbers, one of which would be "NP" for the "number of planets", and suppose "$\Box$ $(9 > 7)$" is true, according to our system. But say we also have "$\sim \Box$ $(NP > 7)$". Now suppose "NP" is taken to be as legitimate a name for the number of planets as "9", (i.e. for this *number*) as the numeral itself. Then we get the odd seeming conclusion, (anyway in your (Marcus's) quantification) that

$$(\exists\ x, y)\ (x = y \cdot \Box\ (x > 7) \cdot \sim \Box\ (y > 7))$$

On the other hand, if "*NP*" is not taken to be as legitimate a name for the number of planets as "9", then, in that case, I presume that Quine would reply that this sort of distinction amounts to the distinction of essentialism itself. (To Quine:) Would this be a good way of stating your position?

Prof. Quine: Yes. And I think this formula is one that Professor Marcus would accept under a new version of quantification. Is that right?

Prof. Marcus: No... this wouldn't be true under my interpretation, if the '=' (of Kripke's expression) is taken as identity. If it were taken as identity, it would be not only odd-seeming but contradictory. If it is taken as '$=_m$' then it is not odd-seeming but true. What we must be clear about is that in the extended modal systems with which we are dealing here, we are working within the framework of the theory of types. On the level of individuals, we have only identity as an equivalence relation. On the level of predicates, or attributes, or classes, or propositions, there are other equivalence relations which are weaker. Now the misleading aspect of your (Kripke's) formulation is that when you say, "let the variables range over the numbers", we seem to be talking about individual variables, '=' must then name the identity relation and we are in a quandary. But within a type framework, if x and y can be replaced by names of numbers, then they are higher type variables and the weaker equivalence relations are appropriate in such contexts.

Mr. Kripke: Well, you're presupposing something like the Frege-Russell definition of number, then?

Prof. Marcus: All right. Suppose numbers are generated as in *Principia* and suppose 'the number of planets' may be properly equated with '9'. The precise nature of this equivalence will of course depend on whether 'the number of planets' is interpreted as a description or a predicate, but in any case, it will be a weak equivalence.

Mr. Kripke: Nine and the number of planets do not in fact turn out to be identically the same?

Prof. Marcus: No, they're not. That's just the point.

Mr. Kripke: Now, do you admit the notion of 'identically the same' at all?

Prof. Marcus: That's a different question. I admit identity on the level of individuals certainly. Nor do I foresee any difficulty in allowing the identity relation to hold for objects named by higher type expressions (except perhaps propositional expressions), other than the ontological consequences discussed in the paper. What I am *not* admitting is that 'identically the same' is indistinguishable from weaker forms of equivalence. It is explicit or implicit extensionalizing principles which obliterate the distinction. On this analysis, we could assert that

9 is identically the same as 9

but not

9 is identically the same as (5 + 4)

without some weak extensionalizing principle which reduces identity to logical equivalence.

Mr. Kripke: Supposing you have any identity, and you have something varying over individuals.

Prof. Marcus: In the theory of types, numbers are values for predicate variables of a kind to which several equivalence relations are proper.

Mr. Kripke: Then, in your opinion the use of *numbers* (rather than individuals) in my example is very important.

Prof. Marcus: It's crucial.

Prof. Quine: That's what I used to think before I discovered the error in Church's criticism. And if I understand you, you're suggesting now what I used to think was necessary; namely, in order to set these things up, we're going to have, as the values of variables, not numbers, but assorted number properties, that are equal, but different – the so-called number of planets on the one hand, 9 on the other. What I say now is that this proliferation of entities isn't going to work. For example, take x as just as narrow and intensional an object as you like...

Prof. Marcus: Yes, but not on the level of individuals where only one equivalence relation is present. (We are omitting here consideration of such relations as congruence.)

Prof. Quine: No, my x isn't an individual. The values of 'x' may be properties, or attributes, or propositions, that is, as intensional as you like. I argue that if $\varphi(x)$ determines x uniquely, and if p is not implied by $\varphi(x)$, still the conjunction $p \cdot \varphi(x)$ will determine that same highly abstract attribute, or whatever it was, uniquely, and yet these two conditions will not be equivalent, and therefore this kind of argument can be repeated for it. My point is, we can't get out of the difficulty by splitting up the entities; we're going to have to get out of it by essentialism. I think essentialism, from the point of view of the modal logician, is something that ought to be welcome. I don't take this as being a *reductio ad absurdum*.

Prof. McCarthy: (MIT) It seems to me you can't get out of the difficulty by making 9 come out to be a class. Even if you admit your individuals to be much more inclusive than numbers. For example, if you let them be truth values. Suppose you take the truth value of the 'number of planets is nine', then this is something which is true, which has the value truth. But you would be in exactly the same situation here. If you carry out the same problem, you will still get something which will be 'there exists x, y such that $x = y$ and it is necessary that x is true, but it is not necessary that y is true'.

Prof. Marcus: In the type framework, the individuals are neither numbers, nor truth values, nor any object named by higher type expressions. Nor are the values of sentential variables truth values. Sentential or propositional variables take as values sentences (statements, names of propositions if you will). As for your example, there is no paradox since your '$=$' would be a material equivalence, and by virtue of the substitution theorem, we could not replace 'y' by 'x' in '$\square$ x' (x being contingently true).

Prof. McCarthy: Then you don't have to split up numbers, regarding them as predicates either, unless you also regard truth functions as predicates.

Prof. Marcus: About "splitting up". If we must talk about objects, then we could say that the objects in the domain of individuals are extensions, and the objects named by higher order expressions are intensions. If one is going to classify objects in terms of the intension – extension dualism, then this is the better way of doing it. It appears to me that a failing of the Carnap approach to such questions and one which generated some of these difficulties, is the passion for symmetry. Every term (or name) must, according to Carnap, have a dual role. To me it seems unnecessary and does proliferate entities unnecessarily. The kind of evidence relevant here is informal. We do, for example, have a certain hesitation about talking of identity of propositions and we do acknowledge a certain difference between talking of identity of attributes as against identity in connection with individuals. And to speak of the intension named by a proper name strikes one immediately as a distortion for the sake of symmetry.

Follesdal: The main question I have to ask relates to your argument against Quine's examples about mathematicians and cyclists. You say

111

that (55) is not provable in $QS4$. Is your answer to Quine that it is not provable?

Prof. Marcus: No. My answer to Quine is that I know of no modal system, extended of course, to include the truth of:

It is necessary that mathematicians are rational

and

It is necessary that cyclists are two-legged

by virtue of meaning postulates or some such, where his argument applies. Surely if the argument was intended as a criticism of modal logic, as it seems to be, he must have had *some* formalization in mind, in which such paradoxes might arise.

Follesdal: It seems to me that the question is not whether the formula is provable, but whether it's a well-formed formula, and whether it's meaningful.

Prof. Marcus: The formula in question is entirely meaningful, well-formed if you like, given appropriate meaning postulates (defining statements) which entail:

All mathematicians are rational

and

All cyclists are two-legged .

I merely indicated that there would be no way of *deriving* from these meaning postulates (or defining statements) as embedded in a modal logic, anything like:

It is necessary that John is rational

given the truth:

John is a mathematician

although both statements are well-formed and the relation between 'mathematician' and 'rational' is analytic. The paradox simply does not arise. What I *did* say is that there is a derivative sense in which one can talk about necessary attributes, in the way that abstraction is derivative.

For example, since it is true that

$$(x) \; \square \; (xIx)$$

which with abstraction gives us

$$(x) \; \square \; (x \epsilon \hat{y}(yIy))$$

which, as we said before, would give us

$$\vdash \square \; \hat{y}(yIy)$$

The property of self-identity may be said to be necessary, for it corresponds to a tautological function. Returning now to Professor Quine's example, if we introduced constants like 'cyclist', 'mathematician', etc., and appropriate meaning postulates then the attribute of being either a non-mathematician or rational, would also be necessary. Necessary attributes would correspond to analytic functions in the broader sense of analytic. These may be thought of as a kind of essential attribute, although necessary attribute is better here. For these are attributes which belong necessarily to every object in the domain whereas the usual meaning of essentialism is more restricted. Attributes like mathematician and cyclist do not correspond to analytic functions.

Prof. Quine: I've never said or, I'm sure, written that essentialism could be proved in any system of modal logic whatever. I've never even meant to suggest that any modal logician even was aware of the essentialism he was committing himself to, even implicitly in the sense of putting it into his axioms. I'm talking about quite another thing – I'm not talking about theorems, I'm talking about truth, I'm talking about true interpretation. And what I have been arguing is that if one is to quantify into modal contexts and one is to interpret these modal contexts in the ordinary modal way and one is to interpret quantification as quantification, not in some quasi-quantificatory way that puts the truth conditions in terms of substitutions of expressions, – then in order to get a coherent interpretation one has got to adopt essentialism, and I already explained a while ago just how that comes about. But I did not say that it could ever be deduced in any of the S-systems or any system I've ever seen.

Prof. Marcus: I was not suggesting that you contended that essentialism could be *proved* in any system of modal logic. But only that I know of no interpreted modal system, even where extended to include predicate constants such as those of your examples, where properties like being a mathematician would necessarily belong to any object. The kind of uses to which *logical* modalities are put have nothing to do with essential properties in the old ontological sense. The introduction of physical modalities would bring us closer to this sort of essentialism.

Follesdal: That's what creates the trouble when one thinks about properties of this kind, like being a cyclist.

Prof. Quine: But then we can't use quantifiers as quantifiers.

Prof. Marcus: The interpretation of quantification has advantages other than those in connection with modalities. For example, many of the perplexities in connection with quantification raised by Strawson in *Introduction to Logical Theory* are clarified by the proposed reading of quantification. Nor is it my conception. One has only to turn to the Introduction of *Principia Mathematica* where existential quantification is discussed in terms of 'always true' and 'sometimes true'. It is a way of looking at quantification that has been neglected. Its neglect is a consequence of the absence of a uniform, colloquial way of translating although we can always find some adequate locution in different classes of cases. It is *easier* to say 'There is a thing which...' and since it is adequate some of the time it has come to be used universally with unfortunate consequences.

Prof. Quine: Well, Frege, who started quantification theory, had the regular ontological interpretation. Whitehead and Russell fouled it up because they confused use and mention.

Follesdal: It seems from the semantical considerations that you have at the end of the paper, that you need your special axiom.

Prof. Marcus: Yes, for that construction. I have no strong preferences. It would depend on the uses to which some particular modal system is to be put.

Follesdal: You think you might have other constructions?

Prof. Marcus: Indeed. Kripke, for example, has suggested other constructions. My use of this particular construction is to suggest that in discussions of the kind we are having here today, and in connection with the type of criticism raised by Professor Quine in *Word and Object* and elsewhere, it is perhaps best carried out with respect to some construction.

Mr. Kripke: Forgetting the example of numbers, and using your interpretation of quantification – (there's nothing seriously wrong with it at all) – does it not require that for any two names, 'A' and 'B', of individuals, 'A = B' should be *necessary*, if true at all? And if 'A' and 'B' are names of the same individual, that any necessary statement containing 'A' should remain necessary if 'A' is replaced by 'B'?

Prof. Marcus: We might want to say that for the sake of clarity and ease of communication that it would be convenient if to each object there were attached a single name. But we can and we do attach more than one name to a single object. We are here talking of proper names in the ideal sense, as tags and not descriptions. Presumably, if a single object had more than one tag, there would be a way of finding out such as having recourse to a dictionary or some analogous inquiry, which would resolve the question as to whether the two tags denote the same thing. If 'Evening Star' and 'Morning Star' are considered to be two proper names for Venus, then finding out that they name the same thing as 'Venus' names is different from finding out what is Venus' mass, or its orbit. It is perhaps admirably flexible, but also very confusing to obliterate the distinction between such linguistic and properly empirical procedures.

Mr. Kripke: That seems to me like a perfectly valid point of view. It seems to me the only thing Professor Quine would be able to say and therefore what he must say, I hope, is that the assumption of a distinction between tags and empirical descriptions, such that the truth-values of identity statements between tags (but not between descriptions) are ascertainable merely by recourse to a dictionary, amounts to essentialism itself. The tags are the "essential" denoting phrases for individuals, but empirical descriptions are not, and thus we look to statements containing "tags", not descriptions, to ascertain the essential properties of individuals. Thus the assumption of a distinction between "names" and "descriptions" is equivalent to essentialism.

Prof. Quine: My answer is that this kind of consideration is not relevant to the problem of essentialism because one doesn't ever need descriptions or proper names. If you have notations consisting of simply propositional functions (that is to say predicates) and quantifiable variables and truth functions, the whole problem remains. The distinction between proper names and descriptions is a red herring. So are the tags. (*Marcus*: Oh, no.)

All it is is a question of open sentences which uniquely determine. We can get this trouble every time as I proved with my completely general argument of p in conjunction with φx where x can be as finely dis-

115

criminated an intension as one pleases – and in this there's no singular term at all except the quantifiable variables or pronouns themselves.

Mr. Kripke: Yes, but you have to allow the writer what she herself says, you see, rather than arguing from the point of view of your own interpretation of the quantifiers.

Prof. Quine: But that changes the subject, doesn't it? I think there are many ways you can interpret modal logic. I think it's been done. Prior has tried it in terms of time and one thing and another. I think any consistent system can be found an intelligible interpretation. What I've been talking about is quantifying, in the quantificational sense of quantification, into modal contexts in a modal sense of modality.

Mr. Kripke: Suppose the assumption in question is right – that every object is associated with a tag, which is either unique or unique up to the fact that substituting one for the other does not change necessities, – is that correct? Now then granted this, why not read "there exists an x such that necessarily p of x" as (put in an ontological way if you like) "there exists an object x with a name a such that p of a is analytic." Once we have this notion of name, it seems unexceptionable.

Prof. Quine: It's not very far from the thing I was urging about certain ways of specifying these objects being by essential attributes and that's the role that you're making your attributes play.

Mr. Kripke: So, as I was saying, such an assumption of names is equivalent to essentialism.

Prof. Cohen: I think this is a good friendly note on which to stop.

5

ALAN ROSS ANDERSON / OMAR KHAYYAM MOORE

MODAL LOGICS II: TOWARD A FORMAL ANALYSIS

OF CULTURAL OBJECTS*

Presented March 8, 1962

In this essay we hope to make some progress toward an explication or rational reconstruction of the concept *culture*, which has been of interest to philosophers at least since the time of Hegel [8], and also to social scientists since the 1870's, when Tylor [25] explicitly introduced the concept as an analytic category. Our aim is not only to clarify the concept, but also to indicate some unfortunate effects of philosophical naïveté on the part of some social scientists, who have in fact been embroiled in problems calling for careful philosophical analysis, without recognizing the importance of philosophical techniques. We hope also to show the relevance of our analysis to some traditional philosophical problems, such as the realist-nominalist ontological debate, and the continuing critique of Aristotelian essentialism. And we remark finally that the problem is by no means artificial: we feel that the issues we hope to clarify have been genuinely perplexing both to philosophers and to social scientists.

As will become clear, the things we construe as cultural are *abstract*, or *conceptual* (depending on whether one wants to take a Platonic or a Kantian approach). As we shall try to point out at the end of the paper, the issues posed here seem to require a much more adequate philosophical analysis than they have yet received.

We hope also to show, in an incidental way, that the developments in modal logic, beginning with C. I. Lewis [13], have a direct bearing on our topic. Since his pioneering work, dozens of systems have been developed, and within the past few years a satisfactory uniform treatment has been

* This research was supported in part by the Office of Naval Research, Group Psychology Branch, Contract $\neq$ SAR/Nonr-609 (16). We are forced again as in our last article [17] to apologize to our friends Jon Barwise, Nuel D. Belnap, Jr., and Neil Gallagher, for boring them to tears in the course of reading and re-reading earlier drafts of this article. They are not to be held accountable for our errors, but they have surely saved us from many.

effected, largely due to the efforts of Kripke [10] and Hintikka (some un-published work). We shall not be concerned to discuss formal questions about such systems, but rather to point out, as we go along, connections between systems having to do with *possibility, permission,* and other modal notions, and the concept *culture.* We ask the reader to take upon himself the task of noting how often we use the modal words "can", "must", etc., in an essential way.

We begin by considering some definitional problems suggested by our title.

I

Attempts at rational reconstructions are in a certain sense quests for definitions, and in view of the positivistic attack on essentialism, we feel called upon at the outset to explain at least some ways in which definitions may be theoretically fruitful. It is well known that there is a sense in which the only terms needed in a theory are the undefined ones – namely, in the sense that all explicitly defined constants in a formal theory are in principle eliminable. But preoccupation with this feature of definitions (as for example by Popper [20]), may very well lead us to overlook cases where the foundation of a theory lies in a fundamental definition.

As an illustration of the point, we select the theory of games. It is true that in spite of the vagueness of the term "game" in ordinary language, it is nevertheless possible for us to recognize many things (such as baseball and chess) as games, and many others (ships, shoes, sealing wax) as nongames. But it is equally true that it would be virtually impossible to construct a decent *theory* of games simply on the basis of our ability to recognize them. What is required is an adequate definition of the notion, and furthermore, a definition with a degree of rigor and precision not to be found in the dictionary. We owe to Borel [5] and Von Neumann [19] a definition of "game" which is sufficiently clear to serve as the basis for a fruitful and interesting theory about games – and there is a sense in which we may regard their definitions as an answer not to the question, "What *do* we mean by 'game'?" but rather to the question "What *ought* we to mean by 'game'?" We can pose the latter question also in the following form: "How can we define 'game' in such a way as (a) to get an interesting formal theory off the ground, while (b) minimizing conflict with informal usage?" In point of fact, the reconstruction or definition offered by Borel and Von

Neumann is sufficiently close to our usual sense of the word, so that chess, checkers, baseball, and many other social games, *do* count as "games" in their technical sense; we are now in a position to characterize and analyze what we may take to be the "essential" features of games as usually understood.

What we would like, of course, would be to find an equally profound definition of "culture" – profound in the sense of leading deductively to interesting and illuminating consequences. This we are at the moment unable to provide. But in what follows we shall at least be able to state some conditions which *any* satisfactory definition of the concept will have to satisfy – and in that way to a certain extent limit the area within which the search should be carried on. We subsequently offer definitions of *culture* and *cultural object* which, while not satisfying our own standards of rigor, seem at least to be steps in the right direction.

II

Conditions of adequacy. One of the first and most obvious of the questions to ask is what we expect the concept of culture to do for us. This question may most easily be asked by considering some statements involving the term "culture," or "cultural object," which we would like the theory to certify as true. Without attempting to make any sort of exhaustive list or classification, we present some examples of statements which any student of the matter would recognize as somehow in the spirit of things.

1. The principal language spoken in Rome has changed substantially between Cicero's time and our own. That is, considering the language as a cultural object, we would want to be able to distinguish between modern Italian and Latin, and describe this difference as an instance of *cultural change*.

2. The game of *morra*, on the other hand, was played in Rome in Caesar's day, but is also played there now. Again, conceiving of the game as a cultural object, we want to be able to discuss the *persistence* of this cultural object in a society.

3. The game of Gō, though it originated in the Orient, and persists there, is now also played in the western hemisphere – in Brooklyn for example – and here we find an instance of *cultural diffusion*, the movement of cultural objects from one society to another.

4. But cultural diffusion is not the only way in which new cultural objects come to a society: cultural objects may be created or rediscovered.

119

Knowledge of how to read the Linear B orthography was for a long time lost, and has only recently been rediscovered. The custom of programming electronic computers was developed only recently, and new forms of the art are continually being elaborated. I.e., we wish to be able to talk about cultural *innovation*, as well as the *disappearance* and *reappearance* of cultural objects in a society.

These are not all of the things we wish to say about cultural objects, but it is evident that we must say some of these things if we are to undertake any kind of comparative studies of cultural objects or systems of cultural objects. We *must* be able to say that the tensor calculus was not a part of classical Greek culture.

All this must sound very elementary, but careful study makes it evident that most of the definitions of "culture" found in standard sources, and even in technical monographs devoted to the topic, explicitly preclude saying just exactly the kind of thing that requires to be said. This is a strong statement, and we now proceed to give it sufficient documentation to convince the reader of its truth.

III

Review of the literature. In reviewing the literature on culture, we have found ourselves in the position of a logician who is forced to read endless books which fail to distinguish between use and mention; or of an analytic philosopher compelled to review endless books in which the task-verb and achievement-verb distinction is overlooked; or of a billiard player condemned to play with a twisted cue, on a cloth untrue, ... The situation taxes one's patience, and if we appear to be polemical, we apologize. To keep the discussion appropriately impersonal, we will worry only about the definitions listed below, leaving authors aside, remarking only that all the definitions we consider may be found in Kroeber and Kluckhohn's [12] careful and exhaustive treatment of the topic.

It will be convenient to classify the definitions of "culture" under a number of headings, according to the kind of galimatias involved.

1. *Complex wholes.* "Complex whole" (a phrase stemming initially from Tylor [25]) crops up in many of the definitions; sometimes in the variant "integral whole." Examples:

Culture, or civilization... is that complex whole which includes knowledge, belief, art, morals, law, custom, and any other capabilities and habits acquired by man as a member of society.
... that complex whole which includes all the habits acquired by man as a member of society.
... obviously is the integral whole consisting of implements and consumer's goods, of constitutional charters for the various social groupings, of human ideas and crafts, beliefs and customs.
Culture is that complex whole which includes artifacts, beliefs, art, all the other habits acquired by man as a member of society, and all products of human activity as determined by these habits.

Aside from some minor perplexities in each of the definitions (the list of cultural items in the third definition suggests that it was produced by a randomizing device, and what kind of good-natured spoofing is going on with the use of "other" in the last definition: "... artifacts, beliefs, art, all the *other* habits..."?), we may note at once that if culture must be regarded as such a monolithic "complex whole", it is difficult to see how cultural diffusion could ever take place, or how new items could enter the system without producing a new system. Some writers have even suggested, if not explicitly said, that a game, for example, must be regarded as *two* cultural objects, if it is played in different societies in different cultural contexts:

The occupation with living cultures has created a stronger interest in the totality of each culture. It is felt more and more that hardly any trait of culture can be understood when taken out of its general setting. [4]

We would not want to define culture in such a way that the study of *systems* of cultural objects was rendered impossible, or in such a way that properties of such systems could not be investigated, but it would seem difficult to discuss systems of cultural objects without some idea of what we want to mean by "*a* cultural object", such as "the language of the Basque", and no such idea is provided by holistic definitions such as these.

Though these definitions *seem* to promise the "broad" character required, they suffer from defects. Putting aside the fact that we do not know very well what the difference between a complex and a simple whole is, or whether a simple whole may have complex parts, or the like – even if we understood these notions it turns out that the definition is too narrow. A society might have a culture which, though complex, was sufficiently disorganized so that we would want to admit almost no integrity. In any event it is surely an empirical question whether a society has a culture consisting of an "integral complex whole". Such notions should not be analytically built into a definition of the concept of culture.

2. *Identification with behavior*. Recognition of the fact that culture has *something* to do with learned behavior has led certain writers to identify culture with learned behavior or with some kind of learned behavior.

... culture is the sociological term for learned behavior...
... culture... may be defined as *all behavior learned by the individual in conformity with a group.*

But on the whole, those who wish to include learned behavior under culture have been sufficiently uneasy about the matter to include also *patterns* or *ways* of behaving, sometimes flatly contradicting themselves (or so it seems to us):

The culture of a society may be said *to consist of* the characteristic *ways* in which basic needs of individuals are satisfied in that society (that is, *to consist of the particular response sequences* of various behavior-families which occur in the society)... [our italics].

In the passage just cited, the author seems to want to make a distinction between particular response sequences, such as that sequence of responses on the part of Hitler which comprised his dancing a jig on the occasion of the fall of France (an episode in Hitler's own psycho-biological history, which can never be transmitted to another – nor can anyone else ever be Hitler dancing the *particular* jig), and "behavior families", which are presumably such things as jig-dancing, which we would hold *is* a transmissible cultural object. But having made the distinction, the writer promptly obliterates it by saying "that is".

We would agree that jig-dancing is a cultural object, as are all other learned dance forms, but it seems equally clear that happenings, or events, or particular bits of behavior, or particular response sequences, are highly perishable, and certainly *not* the kind of thing which is transmissible, persisting, changing through or across generations – in short behavior is *not* the kind of thing which could reasonably be regarded as a cultural object, learned or not.

To take another example, the two-valued propositional calculus certainly ought to rank as a cultural object, since it is the kind of thing which has been passed down through several generations of logicians, which can diffuse, which persists as a stable entity, and which can fit in with other cultural objects (quantification theory, for example, or modal logic) in a systematic way. And of course we would not want to say that the propositional calculus was present in the culture of a society unless at least some people in that society thought about it, taught it, wrote about it, used it, or the like. But it is equally clear that none of these behavings relative to the propositional calculus can be identified with the propositional calculus. Though it may be in some way related to learned behavior, it *is not* learned behavior. It is not behavior of any sort. Neither are board games, nor rules for parking, nor methods for tying knots.

It is rather one of the virtues of the concept *culture* as anthropologists *use* it (rather than define it) that it permits focussing attention on those aspects of human situations which can remain invariant under changes of personnel, equipment, and setting of time and place. And we should point out that there are some anthropologists who seem to appreciate this point, notably Kroeber [11] and White [28].

3. *The social heritage.* In spite of the fact that most social scientists are very much interested in the influence of contemporaries on contemporaries, there are in the literature many definitions of culture, according to which an object is not a cultural object unless it has a long history – so that in effect a society can acquire a cultural object just in case it already has it.

... the socially inherited assemblage of practices and beliefs that determines the textures of our lives...
Culture is thus the *social heritage*, the fund of accumulated knowledge and customs through which the person "inherits" most of his behavior and ideas.
The term *culture* is used to signify the sum-total of human creations, the organized result of human experience up to the present time.

Many of the things we learn were developed in times past, but it is difficult to see why this fact should be taken as in any way essential to the

123

notion of *culture*. It would seem to follow from such definitions that no cultural innovation is possible; one may even be led to wonder how systems of cultural objects ever came into existence.

Counterexamples crowd into the mind. Richard Friedberg's [6] solution to Post's problem [21], discovered while Friedberg was a senior at Harvard, was an innovation which certainly passed quickly from generation to generation – but it went backward; and it is now likely that all those interested in the topic, save perhaps the very oldest, are familiar with his result. Our language unfortunately lacks a familiar word to denote such a backward heritage, but an object subject to such a reverse inheritance should certainly not lose its status as a cultural object, simply because it was discovered or created by a prodigy. We should be able to recognize a cultural object independently of any such temporal relations between its originator(s) and recipients.

Furthermore, it is, from a theoretical point of view, undesirable to define "culture" in such a way as to apply only to an empirically determined collection of cultural objects. If the historical approach implicit in the definitions given above were taken seriously, then the only cultural objects would be those which have in fact been produced. But clearly any satisfactory definition of "cultural object" must be such that we can classify a new entity as being or not being a cultural object; we are interested, that is, in a theory of *possible* cultural objects, which, while it will be relevant to the history of societies, will also enable us to consider what is transmissible, independently of historical circumstances. (Notice, incidentally, that the use of the modal concept of *possibility* is essential in making this point.)

4. *Culture as communal.* Anthropologists have on the whole studied the culture of relatively isolated, autonomous, small societies, and it is perhaps this preoccupation which has led to a number of definitions emphasizing the fact that some systems of cultural objects are shared by many members, sometimes by all members, of a society:

Culture is the sum total of the ways of doing and thinking, past and present, of a social group.
... all the things that a group of people inhabiting a common geographical area do...

124

A culture is a given people's way of life, as distinct from the life-ways of other people.

... the term includes those objects or tools, attitudes, and forms of behavior whose use is sanctioned under given conditions by the members of a particular society.

But emphasis on the fact that cultural objects are widely shared by members of a society makes it awkward to consider the culture of an individual, or of a subcollection of the members of a society.

Consider the following example. In 1916, when Herbert O. Yardley, one of the pioneers in American cryptanalysis, handed his *Solution of American Diplomatic Codes* to the bureau chief, he in effect gave him a cultural object consisting of a very powerful method for approaching cryptographic problems. [29] Yardley reports that after a few days, the person to whom he had given the document called him in for the express purpose of determining that he and Yardley were the only two people who knew even of the existence of Yardley's techniques. And indeed Yardley had told no one else about the matter.

Now it seems clear that methods of breaking codes, like methods for making canoes, or baskets, or methods for performing rain-dances, should be regarded as cultural items, even if, as in this case (and at that time) only one person had learned about it. And the fact that crypt-analytic methods had been developed in Europe earlier is certainly not relevant to the status of Yardley's technique as a cultural item – it would have remained a cultural item even if no one else ever thought of it, and even if Yardley's procedures had not been studied and elaborated by the Cryptographic Bureau, as subsequently happened. Our point is simply that for something to qualify as a cultural object, it need not be shared by a vast collection of people: two is enough.

And the example also serves to bring out a second point, namely, that there is no necessary connection between the general availability of a cultural item in a society, and its importance *vis-a-vis* predicting the subsequent course of events. Even by 1921, when the Cryptographic Bureau of the State Department had reached considerable size, it still constituted a miniscule proportion of the population of the United States. But without taking into account the activities of those few persons dealing with cultural objects of a special and little-known sort, it would be difficult to account for some subsequent international history. Since the Black Chamber (as Yardley calls it) was reading the Japanese diplomatic code during the Washington Naval Disarmament Conference, the government of the United States knew ahead of time the minimum ratio of naval forces that the Japanese government was prepared to accept. So all that the United States had to do was to demand the maximum concession obtainable from the Japanese, and then simply wait until the Japanese capitulated. As Yardley remarks, "Stud poker is not a very difficult game after you see your opponent's hole card."

It would be hard to establish that achievements such as these are part of the American "way of life, as distinct from the life-ways of other people" (to return to one of the quotations above). It is evident that Yardley's novel method should count as a cultural object and, as it developed, an important one, though not widely shared. In fact its importance *depended* on its not being widely shared.

5. *Culture and values.* Most social scientists take some pains when discussing culture, to point out that they do *not* mean "culture" in the sense that Maggie has cultural pretensions, while Jiggs does not, i.e. in one of the senses ascribed to the word by Webster:

The enlightenment and refinement of taste acquired by intellectual and aesthetic training.

And it is perhaps in attempting to make this point that the lists of cultural objects, offered in connection with definitions of the term, so frequently are made up of a motley lot of entities:

The various industries of a people, as well as art, burial customs, etc.
...those beliefs, customs, artistic norms, food-habits, and crafts...
...language, marriage, property system, etiquette, industries, art, etc....

What the writers no doubt mean to suggest is that *The Saints Come Marching In* is just as much a cultural item as is *The St. Matthew Passion.* There are of course exceptions to this rule:

Culture is the dissipation of surplus energy in the exuberant exercise of the *higher* human faculties [our italics].

126

By culture we shall understand the sum of all sublimations, all substitutes, or reaction formations, in short, everything in a society that inhibits impulses or permits their distorted satisfaction.

The second of these accounts is as grim as the first is rosy, but such deviations are rare. There is consensus that cultural items may be either lofty or mundane; neither situation affects status as a cultural item, a point with which we are in complete agreement.

There is, however, one evaluative condition to which, according to many theorists, cultural objects must conform: they must be adaptive, adjustive, or somehow help solve problems. In short, they must work.

Culture consists of traditional ways of solving problems...
Culture... is composed of responses which have been accepted because they have met with success; in brief, culture consists of learned problem-solutions.
The culture of society may be said to consist of the characteristic ways in which basic needs of individuals are satisfied in that society...
In its broadest sense, culture is coterminous with everything that is artificial, useful, and social, employed by man to maintain his equilibrium as a biopsychological organism.
The *culture* of a people may be defined as the sum total of the material and intellectual equipment whereby they satisfy their biological and social needs and adapt themselves to their environment.
Culture always, and necessarily, satisfies basic biological needs and secondary needs derived therefrom.

These definitions are of course vague about what constitutes "satisfying basic needs", "useful to maintain equilibrium", "adapt themselves to their environment", and the like – but it would be reasonable to suppose that in *some* sense a cultural object might be maladaptive, or conducive to the extinction of the society. One thinks immediately of the Shakers, who take vows of celibacy, or the Trappist monks, who take vows of celibacy *and* silence, or of institutionalized suicide, such as Japanese Seppuku.

Of course we realize that a really determined Panglossian could find that suicide is socially adaptive, and that silence caters to *some* interest on the part of the monks; such a Panglossian would be delighted to learn from an ordinary language philosopher that *Seppuku-suru* is a success-verb. But even if he can bring himself to believe that the Shakers are exhibiting "adaptive" behavior by adapting themselves out of existence as a social group, we would want empirical evidence to this effect, and not to have the fact follow from the definition.

But it is hard to see how *anyone* could regard as "adaptive" a culture which led to a nuclear war ending in total extinction of all life forms on the planet. We do not wish to press the argument too hard, but it does seem clear that building into a definition of culture a clause to the effect that it must be adaptive or adjustive, robs us of a distinction which we may plausibly want to apply to different cultural objects – saying that some are adaptive and some are not. It seems equally clear that this is an empirical question, and should not be made analytic of the concept *culture*.

We have thus far been considering what we regard as outright philosophical fumbles, which have in common not only many irrelevant restrictions (such as requiring a "tradition", or insisting that the cultural items be "adaptive", or insisting that they belong to a "people"), but also a certain self-defeating character (as for example requiring that culture be something that can be learned, and then including among cultural items things that *cannot* be learned, such as canoes; or requiring that culture be transmissible, and then including among cultural items things that *cannot* be transmitted, such as a particular response on the part of a particular individual). But there is another common type of error in the literature we are discussing, namely, the following.

6. *The man-made part of the environment.* It is frequently said that culture comprises *everything* that is man-made:

The term *culture* is employed in this book in the sociological sense, signifying anything that is man-made, whether a material object, overt behavior, symbolic behavior, or social organization.
A short and useful definition is: "culture is the man-made part of the environment."
Culture may be regarded as that part of the environment that is the creation of man.
Culture designates those aspects of the total human environment, tangible and intangible, that have been created by men.

128

Culture consists of all products (results) or organismic nongenetic efforts at adjustment.

In the broadest sense culture may mean the sum total of everything which is created or modified by the conscious or unconscious activity of two or more individuals interacting with one another or conditioning one another's behavior.

Now the class of those things which have been influenced by mankind (including of course men themselves) could, if one wished, be singled out for study. But one of the principal motivations in considering cultural items as distinct from other kinds of things, was to exclude from culture such entities as, say, the moon, or carbon, or oxygen, or indeed any of the planets, natural satellites, or elements. But now it develops that we must consider the element einsteinium as a cultural item, owing to the historical circumstance that it is not found in nature, but *has* been made in laboratories. This is sufficiently odd in itself, but leads to even further perplexities. What are we to say about water molecules, which we *can* produce in a laboratory? Does this make water a cultural item after all, or are we to think of only those water molecules that have in fact been produced in a laboratory as cultural items? Of course we *can* understand answers to these questions, and if someone wishes to distinguish those water molecules which have been produced in laboratories from those which have not, we can go along with the distinction; but it seems *prima facie* a fruitless distinction to draw.

And the distinction would have to be drawn in other areas as well. We would have to consider two kinds of lakes – those which are cultural items (being man-made) and those which are not – and similarly for radio waves, light waves, and so on.

Men are affecting the physical environment all the time, and in all kinds of ways, and though it would certainly be reasonable to study relations between men and the physical environment, it seems again strange to insist that anything influenced by man is a cultural item. Why should an insect, accidentally killed by a human being, be considered a cultural object?

The last definition quoted above has at least the merit of requiring human interaction in producing the cultural object, but still allows for the possibility of cultural dead insects, provided only that the insect's death was somehow occasioned by interacting humans, rather than by one person's inadvertence.

129

In a sense, definitions of culture as "the man-made part of the environment", might be regarded as a desperate expedient designed to avoid the clear shortcomings discussed under 1–5 above. So defined, we need no longer worry about whether culture must be traditional, or whether it is appropriate to speak only of the culture of a people, or whether culture must be adaptive, or adjustive, or the like. By considering *all* influences of men on their environments as cultural, these problems *can* be avoided – but such a course is hardly worth the price, since obviously men make or do all kinds of things which anthropologists do not want to regard as cultural, and which as a matter of practice no one ever has regarded as cultural.

IV

At this point (if not before) the reader may well feel that we are being captious and picayune, and taking people literally when they meant only to be understood in a kind of vague and general way. In reply to this charge we have four comments.

In the first place, the way in which these definitions have been presented in the literature indicates that they are *not* offered as vague preliminary approximations, but rather as carefully considered and accurate statements of the intended meaning of the term. And this is certainly the spirit in which they are discussed in the monograph from which we have drawn them. Moreover, with very few exceptions these definitions have been produced by professional social scientists, many of whom are leaders in the field – they are not the offerings of dilettantes.

Secondly, counterexamples, even if they seem farfetched, *are* counterexamples, and should be taken as seriously in discussions of culture as they are in other areas of scientific inquiry. No biologist would tolerate a definition of "vertebrate", according to which none of the things he wants to call "vertebrates" could be so called, or according to which every organism would somehow count as a vertebrate. In those scientific disciplines where some progress has been made in constructing theories, investigators are on the whole careful about how they use words – and it does not seem unreasonable to extend this requirement to the social sciences.

Thirdly, the definitions we have discussed have virtually no points of contact with the on-going research of anthropologists and sociologists. A cursory glance at the literature will reveal dedicated holists discussing techniques of basket-weaving, in spite of their alleged conviction that one cannot discuss cultural objects out of context. In one of the most advanced areas in the social sciences, namely linguistics, investigators have studied languages (which are cultural objects if anything is) in complete abstraction from the extralinguistic context of utterance; it is even allowed that it is possible for an American to speak Swahili. And no one seems to object that the undertaking is founded on sand simply because the language has been taken out of its societal and cultural context.

Finally, the conceptual mistakes on which we have been commenting are remarkably widespread: A careful and conservative check on the one hundred and sixty-four definitions used in our "sample" of definitions reveals a total of at least that many instances of mistakes of the kinds we have discussed. Not one of the one hundred sixty-four met the minimum criteria of adequacy sketched at the outset of our discussion.

Fortunately, most social scientists have kept their theoretical ideas sufficiently isolated from their empirical research, so that empirical studies could be made, and what would have been the hamstringing effect of the definitions, if taken seriously, has in fact been negligible. But this is obviously not a happy situation, nor one conducive to the formulation or testing of interesting theories concerning either behavior or culture.

In spite of these severe criticisms, however, we do not believe that the literature on the topic is *totally* without value. The writers do seem clear, for example, on the point that cultural objects are not inherited genetically, and that whatever the area of their study might be, it is not the same as that of most practicing biologists. In the sequel, we will try to give a less damning interpretation to some of the concepts ingredient to the definitions we have discussed. But this will have to wait on our own positive proposals, to which we now turn.

As a starting point, we return to the conditions of adequacy mentioned earlier. At first glance, it may appear that the cultural items there mentioned – Latin, modern Italian, *morra*, Gō, Linear B orthography, computer programming, and the tensor calculus – have very little in common, and perhaps cannot be conveniently grouped together under *any* heading. But they do have in common one rather striking feature, namely, they are all things that people can learn from each other, a fact which prompts us to define a "cultural object" as a "learnable from" item, that is, a cultural object belongs to the class

$$\hat{\alpha}(\exists x)(\exists y)(x \neq y \text{ and } \diamondsuit (x \text{ learns } \alpha \text{ from } y)).$$

(i.e., belongs to the class of those things α, such that for distinct x and y, it is *possible* that x learns α from y.) (First adumbrations of this idea appear in Moore [15] and Moore and Lewis [18].)

We observe first that, on any reasonable account of what it is "to learn something from someone", the proposal that cultural objects belong to this class does catch a large number of things that anthropologists have wanted to regard as cultural items. In addition to the examples mentioned, we cite methods of counting, raising families and crops, fishing, hunting, singing, dancing, praying, sacrificing. Techniques for all these activities can be learned; and when isolated for study they can be compared, contrasted, and investigated historically. People also learn from others many of the things they believe – that the stars guide our destinies, for example, or that light travels faster than sound. They can also learn that the continuum hypothesis is consistent with the other axioms of set theory, and that the Kwakiutl indulge in pot-latches. It would appear, at least, that many of the objects which have always been recognized as parts of culture are encompassed by this definition.

We have not said that culture is what *is* learned from other persons. So put, the definition acquires an historical or quasi-historical character. But what people have learned or are learning is a matter of historical fact, and what is desired of course is a theory not about what people have in fact learned, but about what they could learn. In game theory, for example, we do not consider what games people actually have played at various times – the theory purports rather to consider all possible games. In the same way, we would like a general characterization of all possible cultural objects.

Even as a preliminary proposal, however, this definition suffers from a number of defects. For one thing, it is couched in terms of "x learns α from y". From our standpoint, the notion of learning something from a person is obscure – it is difficult to know how to answer even the most elementary questions about it. For example, we have no principles of individuation: Has a person who knows the English alphabet learned *one* thing, or twenty-six, or perhaps fifty-two? When an engineer learns how to operate a computer, how many things does he learn? How much of what he learns is learned *from* another human being (perhaps by way of an instruction manual), and how much does he "figure out for himself?" Even in the case of those cultural objects for which we have reasonably stable criteria of identity, the question is hard to answer: does a person who knows two equivalent definitions of a Boolean ring know two things or one?

But the lack of an adequate theory of learning need not preclude further discussion of the definition, as is evident from the following methodological point (which is not original with us). In the course of a theoretical investigation, it is sometimes useful, when we come across a problem which is recalcitrant, simply to assume that it has been solved, and to consider the consequences of having a solution. An example is provided by investigations concerning the consistency of elementary number theory. It was the proposal of Hilbert [9] to *prove* the consistency of this theory by finitary means, a problem which turned out to be refractory. But without a proof of consistency, it was still possible for Gödel [7] to consider the consequences of having a consistency proof of the type Hilbert envisaged – and to deduce from that assumption the startling conclusion that the formalized theory of elementary arithmetic would have to be inconsistent.

We have, unfortunately, no such monumental results in view, but we cite the example in order to show that we need not always remain stuck at the same point. If a problem is unsolved we can simply assume that there is a solution and then continue from that point. And what we propose to do here is to assume that an adequate theory of learning exists, so that we can direct attention to other problems concerning the class of learnable things – i.e. the class of cultural objects in our sense. But we should point out that even in the absence of a satisfactory learning theory, cultural objects thus construed are in *some* cases, at least, clearly identifiable; that is, the distinction between cultural and noncultural objects can in many cases be quite sharply drawn.

A more serious limitation, from an explanatory point of view, is that the definition is in a certain sense extrinsic to the objects under study: The definition tells us nothing about formal or structural properties of cultural objects. It is as if we defined mathematics as the kind of topic which mathematicians discuss professionally. This may be accurate, in the sense that all mathematical topics and no others are caught by the definition, but it tells us nothing about mathematics, and in this sense is extrinsic to mathematics itself.

Some cultural objects have been characterized intrinsically – for example, techniques of counting, structural properties of which are characterized categorically by Peano's axioms. And in consequence of having an intrinsic characterization, we have gained a vast amount of information about the system of natural numbers. More recently, structural properties of games of strategy have been analyzed with clarity sufficient to produce interesting theoretical results. But it is evident that in describing cultural objects as the kind of thing that can be learned, we have a characterization which admits of only minimal deductive consequences (though, as we hope to show, some of these are of more interest than one might at first think).

For all we know, however, this definition may turn out to be less extrinsic than it now seems, depending on how problems in learning theory might be solved. We may envisage a theory of learning which would shed a great deal of light on the *kinds* of things that could be learned; enough, perhaps, to provide the required intrinsic characterization of cultural objects. But this is at the moment a subject only for speculation.

In this connection we should also observe that defining cultural objects as "learnables" does *not* commit us to the view that learning processes are somehow more fundamental, or that social sciences "ultimately reduce" to psychology (whatever that may mean). It may well be that the most fruitful approach is to *begin* with "cultural objects" as primitive in the theory, and to describe *learning* as (vaguely put) a thing that people do with cultural objects. From this perspective, analysis of cultural objects may shed light on learning processes. (We have already taken some steps in this direction in [17], where, by working backward from what people *can* learn, *we* learn something about the structure of personality.) It surely does not seem unreasonable that in attempting to construct a theory of learning, one should pay attention to the sorts of things people actually do learn. For the moment we wish only to point out that, for reasons we shall now take up, it seems desirable to have the class of cultural objects be coextensive with the class of objects which people can learn from others.

In spite of the reservations we have expressed, so construing cultural objects seems to have a certain heuristic value, and as a preliminary to a discussion of some of the points it raises, we will now try to formulate it a little more precisely.

As a paradigm for further discussion we will consider the propositional function

$$x \text{ learns } \alpha \text{ from } y,$$

concerning the interpretation of which we offer the following comments.

For the present, at least, we wish to understand the variables x, $y, \ldots$, as ranging over human beings, and the variables α, $\beta, \ldots$, as ranging over the things people learn. There is no reason in principle why we should not discuss possible non-human cultures; there seems to be some evidence of learning on the part of other biological organisms, and for all we know we may some day want to speak of computers as "learning" from humans, or from other computers. But for present expository purposes, we wish to restrict consideration to human beings.

Nor need we, in principle, require that the variables $x, y, \ldots$, range over individual human beings; again for all we know a satisfactory learning theory might require us to speak of group learning, in some sense which does not reduce simply to the learning of individuals in the group. (We notice that in game theory it is not always the case that a "player" is an individual: bridge is a two-player game. Whether or not the players are individuals is irrelevant to the theory – and some aspects of a satisfactory learning theory *might* provide parallels.) But our discussion will in any event not be sufficiently precise to warrant attempts at generalization in this direction.

Now if we take as an example a task as complicated as learning to read English orthography, or, say, learning to write acceptable English prose, it is apparent that in the general case many individuals are involved in the process of teaching a particular person how to cope with the task. So that we will want "learns from" to be understood in the sense of "learns, perhaps in part, from" so as to allow us to say that a child "learns English from" all the people with whom he speaks in the course of coming to master the language. But this will not preclude our saying that the use of some particular word in English was learned by x from a particular person y. It will in general depend on the analysis of the cultural item α whether or not it will make sense to say that a person learns a specifiable

135

part of α from one person, and another specifiable part from another person. In the case of languages it might make sense so to speak; but it would probably have been impossible for Paderewski, for example, to have isolated the contributions made by Leschetizky, on the one hand, and Raguski on the other (though he did say, "he [Leschetizky] taught me more in those few lessons than I had learned during the whole twenty-four years preceding that time").

Of course we want to allow also for learning from other persons *via* books, or indeed any medium of communication, so that we can learn not only from Plato, but also indirectly from Socrates. But we would probably want to distinguish learning *from* a person, even in this remote way, from learning *because* of another person's activity. (For an analysis of the distinction between "learning from" and "learning because an appropriate environment is provided", see Moore.[16]) Paderewski's parents provided teachers, and an environment suitable for learning to play the piano, but in a reasonably straightforward sense it is clear that the parents did not teach him the piano. In order for us to say that x learns α from y, we require that y know something about α, though we do not require that y actually give formal instruction in α; he may be teaching α to x without realizing he is doing so – teaching by example, say. Mothers do not teach their children to talk by giving them a series of lectures.

The messiness of all these details, and the lack of any satisfactory theory concerning them, is one of the methodological reasons why it seems a good strategy to consider cultural objects in abstraction from the particular users or learners of these objects. Doing so enables us then to define the "culture" of an individual x as the class of those things he has learned from others:

$$\text{Culture of } x =_{\text{def}} \hat{\alpha}(\exists y) [x \text{ learns } \alpha \text{ from } y, \text{ and } x \neq y].$$

And similarly, perhaps more naturally from the usual anthropological standpoint, we can define one possible sense of "the culture of a society A" as the set of things learned from someone by some member of A:

$$\text{Culture of } A =_{\text{def}} \hat{\alpha}(\exists x)(\exists y) [x \text{ is a member of A and } x \text{ learns } \alpha \text{ from } y, \text{ and } x \neq y].$$

Here y may also be a member of A, in which case we would have y contributing to x's "social heritage"; or if y is not in A, then we have a case of cultural diffusion.

136

An alternative definition, which would perhaps capture the notion of "sharing" of cultural objects, would be expressed by the formulation

Culture of $=$ def A $\hat{\alpha}(x)$ [if x is a member of A, then $((\exists y)x$ learns α from y, and $x \neq y)$].

This latter definition would exclude the case of cryptanalysis mentioned earlier.

In some cases, and in many of the cases represented in the literature, the α in question is fairly widespread in the society, but the culture of a society A as first construed above will allow also for the possibility of having in a society, a cultural item which has been learned by very few members. So that definition does encompass the kinds of learnable things which constitute our "social heritage", as well as less well-known items. We may also consider *systems* of cultural objects if we like (perhaps under the heading "complex wholes"), though again no theory is available. Some of the writers previously discussed are interested in even broader classes of cultural objects; e.g., the culture of *mankind*. And here again none of the foregoing militates against considering such a topic and we may, if we like, define human "culture", without further qualifying phrases, as

Culture $=$ def $\hat{\alpha}(\exists x)(\exists y)$ [x learns α from y, and $x \neq y$].

At this point we would like to stand back for a moment and make a methodological observation. In spite of the fact that we have no criteria for measuring the "abstractness" of a notion, it seems natural to remark that "culture" defined in this way is a very abstract concept. We wish to consider things that are learned (or more generally, are learnable) in complete abstraction from learners, teachers, spatio-temporal considerations, motivation, context, and so on. Though it may seem that such abstract considerations are leading away from the empirical domain in which they are supposed to have applications, according to our reading of the history of science some such step is almost always essential to progress in a field. Such notions as mass, temperature, and the like, are *not* tied closely to particular physical objects or particular kinds of physical objects – we do not speak continually of the specific gravity of lead, as opposed to the specific gravity of iron, though of course we *may* do so if we wish. Similarly, it seems unreasonable to insist that we *always* speak of

137

the culture *of* a particular society or people, though again the considerations we have been urging would allow us to do so if we wish.

It may seem that our arguments to the effect that we should restrict "cultural *object*" to the kind of thing that can be learned (propositions, techniques, values, rules, and the like), would prevent us from giving any account of cultural institutions such as corporations, marriages, or seal-hunting among the Eskimos (as opposed to techniques for carrying on as a corporate body, or conducting a marriage, or catching a seal). Or that we could not find any cultural place at all for articles of incorporation, marriage licenses, or kayaks. We reply that our insistence on this restriction, and on the use of the word "object", in the phrase "cultural object", has two points:

(1) In restricting cultural objects to things that are learnable, we mean to be emphasizing the *Reality* of propositions, etc., as *things* or *objects* that can be learned. Our natural and normal ways of talking and thinking demand that we recognize *rules* as such: they may be followed, they may be the topic of a conversation, they may be broken. We mean only to point out that if the notion of culture is to be taken seriously at all, then these abstract entities are at the heart of the matter.

(2) But we are still in a position to distinguish between the cultural objects, in our sense, and products which can be defined in terms of them. Rules for running corporations enable us to run corporations; social rules for running marriages (monogamous, polygamous, polyandrous, whatever you like) enable us to carry on as required; and methods for catching seals enable us, with luck, to catch some seals.

Similarly: Rules for forming corporations enable us to form corporations (and there would be no way of doing it without the rules). Social conventions (of whatever form) enable us to recognize a marriage as such. Rules for application of the term "kayak" enable us to distinguish a kayak from other kinds of boats.

All these institutions and material objects are cultur*al*, in the sense that they are intimately tied (in a variety of ways which no one has spelled out clearly) to cultural objects as we conceive them. So we can admit that there are objects which are closely related to culture, but which *are not* cultural objects in the sense required by the conditions of adequacy we mentioned originally. (And we should point out that we are not ourselves responsible for the conditions of adequacy – they were drawn from the literature. Nor do we feel at the moment called on to defend them – if the reader is not in sympathy with the enterprise anthropologists have set for themselves, then he should have stopped reading long ago.)

To take one of the adequacy criteria:

A corporation cannot diffuse (though rules for forming them can); a marriage cannot diffuse (though the institution of marriage can); and a kayak cannot diffuse (though a method of making them can).

Our point is rather that *if* one thinks that there are important elements in human society which are not adequately covered by those who study physics, chemistry, biology, or psychology – namely, elements which are essentially cultural – then the point on which to concentrate is what we can learn from each other. If our learnings lead to material or institutional *products* of interest, we have no complaint. (In fact we would hope that *our* cultural objects *would* lead to such products.) But the products are not the things learned.

Since we have very little more to say about culture, as regards definitional problems, it may be well to add a few further remarks in defense of the abstractness of the approach by considering some cases where careful and exhaustive analyses of cultural objects have been achieved.

Among the social sciences, the only detailed comparative analyses of really complicated cultural objects are provided by linguists. In addition to the fact that linguists study languages effectively in abstraction from the speakers, number of speakers, etc., they also use in their analyses concepts (such as phoneme, morpheme, generative grammar) which themselves are of a very high level of abstraction. It took linguists a considerable period of time, and much discussion among themselves, to come to see that the phoneme is *not* a particular sound, let alone a sound made by a particular speaker, nor a bit of behavior, nor a response, but rather an entity in abstraction from all of these accoustical happenings, a *property* that some sounds have, relative to a language. And though the set of phonemes of a given language constitutes a system, a "complex whole" if one wishes, it is precisely the kind of "complex whole" which can be diffused from society to society. The analytic categories used by linguists are intended to be applicable to any possible spoken language. In this respect they find their justification in the fact that they work: it *is* possible to compare and contrast different languages conceived of as cultural objects. It is precisely their abstractness and generality which permit such historical investigations as issue in comparisons of Latin with Modern Italian, and thus enable us to characterize certain cultural changes.

Looking at the matter in this way enables us also to see that it is not social scientists alone who have been struggling to characterize and analyze cultural objects. As we remarked before, formal mathematical development of elementary number theory may be regarded as an unusually successful analysis of the process of counting, again considered in abstraction from counters, or from the uses to which counting is put.

Game theorists also have been engaged in analyzing certain cultural objects. And now that we *have* a way of classifying games, we also have the possibility of doing cross-societal studies, which may clarify some perennial theoretical questions raised by anthropologists, philosophers, and others. We have in mind such questions as "Are there any cultural universals?" This question ordinarily receives treatment of two kinds. One common move is to answer with an immediate yes, since all societies have families (if the term is suitably defined), all societies have languages, and so on. But then one has the uneasy feeling that this was after all not quite the answer one wanted – one had been hoping for something with more bite. So the next step is to try the other way of dealing with the question, namely, to restrict the kind of thing that is going to count as a candidate for a cultural universal, so that all of a sudden there are none. In our opinion most of the havering on this issue stems from the lack of suitable analyses of cultural objects. Where analytic techniques are available, as in game theory, we would hope that they would receive some application: we would conjecture that there are some interesting relations between the games played in a society and other features of its system of cultural objects. [22, 23, 24]

In brief, the game theoretical analysis enables us to ask interesting and nontrivial questions which are intermediate between banal questions such as "Are games played in all societies?" (yes), and trivial questions like "Is chess played everywhere?" (no). What we *can* ask is "How are games, falling into various classes depending on their structural properties, distributed throughout societies?" For example "Does *every* society have at least one two-player, zero-sum, generally strictly determined game?"

Notice that when culture is treated in the abstract way we have emphasized earlier, having to do only with what *can* be learned, leaving aside game players, mathematicians, puzzle-solvers, and other such frivolous characters, one has an object with the degree of simplicity requisite for formal, mathematical analysis. But of course such analyses should not remain distant from the empirical world – we always (as we said before, and in other contexts [17] as well) want to see to it that there is a close connection between our formal results and the empirical world in which we all live.

V

It should be immediately obvious, by this time, that in accepting the function $\hat{\alpha}(\exists x)(\exists y)$ [x learns α from y, and $x \neq y$], we are explicitly committed to the existence of abstract entities, directly in opposition to recent nominalistic views. The abstraction operator makes it clear that we are dealing with classes, and the content of the rest of the expression makes it equally clear that the members of the class are abstract entities, e.g., rules of grammar, rules for *bel canto*, rules for constructive proofs in number theory, and so on. This is not the place to go into the question as to the sense in which such entities exist (we have done that elsewhere – see Anderson [1, 2, 3]). But if we are right, it should be apparent that we must learn to live intimately and on a solid intellectual footing with a vast variety of abstract entities, including systems of rules.

It might be debatable whether or not numbers or classes are abstract entities essential to the consideration of problems in anthropology, but there can be absolutely no doubt that *rules*, and *sets of rules*, are of fundamental importance to the social sciences. We have yet to see any satisfactory logical analyses (for unsatisfactory analyses see Mally [14] and Waddell [27]) of rules which do not invoke modal notions. It seems obvious that whatever one takes as arguments in the function "Op" ("It ought to be that p"), the *ought* involved must be treated as a modal operator (cf. Von Wright [26]).

Now we are sympathetic with those who have felt a certain skittishness about the fact that so *many* systems of modal logic have been produced. It almost looks as if we are in a museum: here is a painting, and here is another one, and here is another one, and .... It seems clear that this way of looking at the problem, even if it should yield some light, is not altogether satisfactory. And we can well appreciate the feelings of those who decide

141

that without a uniform treatment of modal systems, nothing much can come of this.

But in the presence of a unified treatment of modal logics, such as those of Kripke, these objections to a serious consideration of modality vanish, and we should now be able to take an attitude toward these problems which will allow us to *use* these conceptual tools. The conceptual tools are available from a number of quarters, and they seem to us to be germane, indeed *necessary*, for any adequate formal analysis of many kinds of cultural objects.

Yale University, New Haven, Connecticut

REFERENCES

1. Anderson, A. R., The Logic of Norms, *Logique et Analyse* **1** (1958).
2. Anderson, A. R., Church on Ontological Commitment, *Journal of Philosophy* **56** (1959) 448–452.
3. Anderson, A. R., What do Symbols Symbolize? Platonism, *Proceedings of the Delaware Conference on Philosophy of Science*. Forthcoming 1963.
4. Boas, F., *Introduction to 'Patterns of culture' by Ruth Benedict*. New York, 1946.
5. Borel, E., The Theory of Play, and Integral Equations with Skew Symmetrical Kernels (Translated by L. J. Savage), *Econometrica* **21** (1953) 97–117.
6. Friedberg, R. M., Two Recursively Enumerable Sets of Incomparable Degrees of Unsolvability (solution of Post's problem, 1944). *Proceedings of National Academy of Sciences* **43** (1957) 236–238.
7. Gödel, K., Über formal unentscheidbare Sätze der Principia Mathematica und verwandter Systeme I, *Monatshefte für Mathematik und Physik* **38** (1931) 173–198.
8. Hegel, G. W. F., *Naturrecht und Staatswissenschaft*, Berlin, 1821.
9. Hilbert, D. and P. Bernays, *Grundlagen der Mathematik*, Berlin, 1934, 1939.
10. Kripke, S. A., A Completeness Theorem in Modal Logic, *The Journal of Symbolic Logic*, **24** (1959) 1–14.
11. Kroeber, A. L., White's View of Culture, *American Anthropologist* **50** (1948) 405–415.
12. Kroeber, A. L. and C. Kluckhohn, *Culture: A Critical Review of Concepts and Definitions*. Papers of the Peabody Museum, Cambridge, 1952.
13. Lewis, C. I., *A Survey of Symbolic Logic*, California, 1918.
14. Mally, E., *Grundgesetze des Sollens*. Graz, 1926.
15. Moore, O. K., Nominal Definitions of Culture. *Philosophy of Science* **19** (1952) 245–256.
16. Moore, O. K., *Automated Responsive Environments: An Application of Sociology to the Problem of Designing Optimal Environment for Learning Complex Cognitive Skills*. National Educational Association Monograph. Forthcoming, 1962.
17. Moore, O. K. and A. R. Anderson, The Structure of Personality, *Journal of Metaphysics*. Forthcoming, 1962.

18. Moore, O. K. and D. J. Lewis, Learning Theory and Culture, *Psychological Review* **59** (1952) 380–388.
19. Neumann, J. von, Zur Theorie der Gesellschaftsspiele, *Math. Annalen* **100** (1928) 295–320.
20. Popper, K. R., *The Open Society and Its Enemies*, London, 1945.
21. Post, E. M., Recursively Enumerable Sets of Positive Integers and Their Decision Problems, *Bulletin of the American Mathematics Society* **50** (1944) 284–316.
22. Roberts, J. M., M. J. Arth, and R. R. Bush, Games in Culture, *American Anthropologist* **66** (1959) 597–605.
23. Roberts, J. M. and B. Sutton-Smith, Child Training and Game Involvement, *Ethnology* **1** (1962) 166–185.
24. Roberts, J. M. and B. Sutton-Smith, Rubrics of Competetive Behavior, *Journal of Genetic Psychology*. Fortcoming 1962.
25. Tylor, E. B., *Primitive culture*, London, 1871.
26. Wright, G. H. von, *An Essay in Modal Logic*, Amsterdam, 1951.
27. Waddell, W., Jr., *Structure of Laws as Represented by Symbolic Methods*, San Diego, 1961.
28. White, L. A., *The science of culture*, New York, 1949.
29. Yardley, H. O., *The American Black Chamber*, Indianapolis, 1931.

RALPH SCHILLER

DETERMINISTIC INTERPRETATIONS OF THE QUANTUM THEORY

Presented March 27, 1962

In this paper we survey several deterministic interpretations of the quantum theory.

Even a superficial analysis of the evolution of our physical ideas reveals no fixed pattern for the creation of successful physical theories. The reasons for the acceptance of a given physical theory by the scientific community are varied, and do not rest solely on the simple notion of agreement with experiment.

As a case in point, we cite Einstein's general theory of relativity. Up until the present moment, there have been three distinct experimental checks of Einstein's theory. Other effects have been predicted by the theory, but they are yet to be observed because of the present unavailability of the sensitive instruments needed for their detection.

In view of the paucity of confirmatory data, it is clear that, in principle, many widely differing theories could be constructed to fit the available evidence. And indeed, several different theories are extant, each in agreement with the known experiments, although, in the main, they ignore the analysis of space-time structure which lies at the heart of Einstein's theory.

In spite of this apparent competition amongst these various theories of gravitation, physicists generally accept Einstein's original version of the theory. Such factors as logical structure, aesthetic appeal, historical antecedence, and the broad predictive powers of Einstein's equations must each play a role in making for so widespread a preference.

Let us accept the validity of Einstein's equations. There still remains the matter of the physical interpretation of these equations, and in this area there are serious differences of opinion. For Einstein, the theory of gravitation was a theory of relativity, a theory in which the laws of nature are to have exactly the same form for all observers, i.e., the laws are to retain their form under arbitrary transformations of the space-time coordinates. These laws of transformation are more general than the laws he discovered in the special (restricted) theory of relativity; hence the name general relativity. And this requirement that a theory of gravitation be at the same time a theory whose form is independent of the observer played a most important part in his discovery of the correct equations for the gravitational field. However, capable investigators have questioned Einstein's view of his theory. I quote one dissenter, V. Fock, to show the seriousness of the controversy over the importance of relativity considerations in gravitational theory. "Enough has been said to make clear that the use of the terms "general relativity", "general theory of relativity" or "general principle of relativity" should not be admitted. This usage not only leads to misunderstanding, but also reflects an incorrect understanding of the theory itself. However paradoxical this may seem, Einstein, himself the author of the theory, showed such a lack of understanding when he named his theory and his publications and when in his discussions he stressed the word "general relativity", not seeing that the new theory he had created, when considered as a generalization of the old, generalizes not the notion of relativity but other, geometrical, concepts."[1]

It is not my purpose to engage in this controversy in relativity theory, or if one prefers, gravitational theory. I have only raised these matters to indicate that questions of interpretation of a physical theory are not peculiar to the quantum theory. I am convinced that, to one degree or another, these problems of interpretation arise in every area of theoretical science.

We turn now to the subject under discussion, the interpretation of the quantum theory. Unlike the experimental situation for gravitational phenomena, the range of application of the quantum theory has proved enormous, and the innumerable experiments confirming the theory have led to the present overwhelming belief in the correctness of the Schrödinger equation and its generalizations, at least in the domain of atomic and electromagnetic phenomena. Coincident with this faith in the equations of the quantum theory has been the general acceptance of the physical interpretation given the theory by Bohr, Heisenberg, and their collaborators, the so-called Copenhagen or orthodox interpretation of the quantum theory.

For most physicists who agree with the Copenhagen school, acceptance of this interpretation of the quantum theory signifies a belief that the laws of nature for microscopic objects are statistical laws, as contrasted with the laws of macroscopic physics which are deterministic. For Bohr, this simple statement of belief does not go far enough, for he feels that the Copenhagen interpretation of the quantum theory has a general philosophical significance, which make such concepts as complementarity applicable to other fields of knowledge. In this matter of extrapolating the concepts of the quantum theory to totally different fields, I believe that one should show the caution adopted by Einstein in similar circumstances when he jokingly warned about "telling the same joke twice".

In reading Bohr and his collaborators, one also gains the impression that it is their belief that the Copenhagen interpretation is the only possible consistent interpretation of the quantum theory. Thus, L. Rosenfeld claims that "if the theory is any good the physical meaning which can be attached to the mathematical equations of the theory is *unique*". [2] This perhaps desirable state of affairs has not held true in other areas of science and there appears no reason to presume that it is valid in the quantum theory. Even if we agreed in principle with Professor Rosenfeld, the meaning of "good theory" would be the subject of endless controversy, and the claim of good theory could hardly deter the critics of the Copenhagen interpretation.

A theory with a unique interpretation possesses the qualities of an ultimate theory, and I do not believe that such a uniqueness claim can be pressed for either the theory itself, or for a specific physical interpretation given the mathematical elements of the theory.

If there is no way of limiting the number of interpretations of a mathematical formalism, this should hardly imply that all such interpretations have equal significance for science. On the contrary, very few (and usually only one) of the alternative interpretations gain widespread acceptability, attesting to a built-in mechanism in the "scientific method" for the defense of the rationality of science, and also to a certain dogmatic element in scientific thought.

As is well known, the Copenhagen interpretation was opposed from the very first by a group of prominent physicists who had provided many of the key ideas in the theory-building process which ultimately led to the present-day quantum theory. This group included Einstein,[3] Schrödinger,[4] Planck,[5] De Broglie,[6] and Von Laue.[7] Their objections eventually were extended to the quantum theory as a physical theory, and not simply to its interpretation. The opposition held, contrary to the Copenhagen school, that either the quantum theory could be understood as a deterministic wave theory, or that the theory was incomplete, for it gave no answer to questions which they felt a good physical theory could be legitimately called on to answer. These questions generally concerned the apparent impossibility of obtaining in quantum processes a classical understanding of individual events. Einstein objected so strongly to the quantum theory that he turned his back on it for the remainder of his life, except for those few occasions when he felt he had found some logical inconsistency in the theory.

Einstein was repelled by the statistical aspects of the quantum theory, for he firmly held to classical determinism, and he had faith that the ultimate theory describing all fundamental processes would have the form of a deterministic field theory modelled after the general theory of relativity. It was his great hope that such a theory, when formulated properly, would retain classical determinism, and in some manner simultaneously yield the statistical assertions of the quantum theory. It is quite possible that Einstein's profound intuition will ultimately prove correct, but, to date, no one has been able to fashion a theory with the qualities sketched above.

De Broglie is motivated by the same distaste for the statistical aspects of the quantum theory, but unlike Einstein, he accepts the validity of the Schrödinger equation. He attempts to introduce determinism in the quantum theory by means of the idea of the double solution. He assumes that for every known continuous solution of the Schrödinger equation, there is another discontinuous solution of the same equation. The continuous solution (the usual solution of the Schrödinger equation) describes a statistical ensemble of particles but this solution is "merely a fictitious wave-function of subjective character, capable only of giving us information of a statistical order about the various possible motions of the particle". On the other hand, the singular solution is to be the "true representation of the physical entity 'particle' which would be an extended wave phenomenon centered around a point (the singular point of the solution), which would constitute a particle in the strictest sense of the word".

De Broglie's scheme might appear to solve the problem of determinism in the quantum theory, but many difficulties have arisen in giving substance to this idea of the double solution, and at the present time it cannot claim to be a completed theory.

It is my feeling that De Broglie's theory, although intuitively elegant, is too closely tied to traditional quantum mechanics for him to fashion a convincing deterministic theory of individual particle motions.

A different approach has been taken by Bohm[8], who proposes that the statistical features of the quantum theory are due to Brownian-type motions which take place on a sub-microscopic level. These motions then give rise to all the observed effects on the microscopic level, for the micro-objects are compounded from the sub-microscopic motions, and *must* obey the Schrödinger equation. To give flesh to such a program is an extremely difficult matter, and at the present time one can hardly say that this program has been fulfilled. Although Bohm's view is pictorial, his assumption of the existence of stochastic variables at a sub-microscopic level seems to me to shift the statistical elements of the quantum theory to an area less accessible to experiment and to physical intuition.

I propose to discuss an alternative deterministic interpretation of the quantum theory by first relating a parable. I want to tell the story of geometrical optics and its evolution as a physical theory.

As is well-known, in geometrical optics the path of a light ray is determined by Fermat's principle which holds that, in travelling from one point in space to another, the ray will take that path which accomplishes the motion in the least time.[9] From this principle, and a knowledge of the optical properties of the medium in which the ray moves, one can show that the rays obey a second order ordinary differential equation closely resembling Newton's law of motion for a particle in a given field of force. Geometrical optics is then deterministic in exactly the same sense as Newton's equations, for if we know the position and velocity of the ray at one point on its path, the remainder of its trajectory can then be determined.

These equations for the rays may be put in the so-called canonical form by introducing an appropriate Hamiltonian,

$$H = \tfrac{1}{2} (p^2 - \mu^2) = 0 \tag{1}$$

where p is the momentum of the ray and $\mu(x)$ the optical index of refraction of the medium.

The canonical equations associated with (1) represent the physics of optics of the 17th and 18th centuries.[10] It is a fully deterministic theory.

At the beginning of the 19th century, a remarkable discovery was made by Thomas Young. If the rays are passed through a narrow slit, they no longer behave as rays; they are diffracted and exhibit typical wave-like interference properties. It was then discovered that the interference effects could be explained by discarding the ray theory and assuming that the rays obeyed the wave equation

$$\nabla^2 \phi + k^2 \mu^2 \phi = 0, \tag{2}$$

where k is proportional to the frequency of wave oscillation. k is an important parameter since it serves to distinguish geometrical optics from the wave theory.

It is clear that our matter (the optical rays or waves) possesses a remarkable duality of character, as it sometimes exhibits the properties of waves, at other times that of particles.[11] Now it is obvious that a thing cannot be a form of wave motion and composed of rays at the same time – the two concepts are too different. It is true that it might be postulated that two separate entities, one having all the properties of a ray, and the other all the properties of wave motion, were combined in some way to form "light". However, such theories are unable to bring about the intimate relation between the two entities which seem required by the experimental evidence. As a matter of fact, it is experimentally certain only that light sometimes behaves as if it possessed some of the attributes of a ray, but there is no experimental proof which proves that it possesses all the properties of a ray; similar statements hold for wave motion. The solution of the difficulty is that the two mental pictures which experiments lead us to form – the one of rays, the other waves – are both incomplete and have only the validity of analogies which are accurate only in limiting cases. It is trite saying that "analogies cannot be pushed too far", yet they may be justifiably used to describe things for which our language has no words, and the apparent duality arises in the limitations of our language. It is not surprising that our language should be incapable of describing processes occurring within distances of 10^{-4} cm, and time intervals of 10^{-14} sec., for, as has been remarked, it was invented to describe the experiences of daily life, and these consist only of processes

149

involving much greater distance and time intervals. Furthermore, it is very difficult to modify our language so that it will describe these processes in the small, for words can only describe things of which we can form mental pictures, and this ability, too, is a result of daily experience. Fortunately, mathematics is not subject to this limitation, and it has been possible to invent a mathematical scheme, the optical duality theory, which seems entirely adequate for the treatment of these processes.

Such a mathematical scheme can be developed by introducing a formalism in which the kinematical and dynamical variables of the classical ray theory are replaced by symbols subjected to a non-commutative algebra. Notwithstanding the renunciation of orbital pictures of rays, Hamilton's equations for classical geometrical optics are kept unaltered, although the conjugate variables x and p of the ray theory now satisfy the rules of commutation,

$$xp - px = i/k . \tag{3}$$

k is the important parameter appearing in the wave equation (2).[12] As a particular representation of the operators x and p of the optical duality theory, we have

$$x_{0p} \to x, p_{0p} \to -\frac{i}{k}\frac{\partial}{\partial x}, \tag{4}$$

and if one assumes that these operators act on a "wave function" ϕ, the classical relations (1) are replaced by the wave equation (2).

The commutation relations (3) show that the knowledge obtainable of the state of an optical system will always involve a peculiar "indeterminacy", for the relation (3) imposes a reciprocal limitation on the fixation of the two conjugate variables x and p, expressed by the "uncertainty" relation,

$$\Delta x \cdot \Delta p \sim \frac{1}{k}, \tag{5}$$

where Δx and Δp are suitably defined latitudes in the determination of these variables. In the limit, as $1/k \to 0$, our optical duality theory goes over into the classical theory of rays, for our non-commuting operators then become the commuting variables x and p of the classical theory, and there is no longer any uncertainty in our description.

If we identify k with $1/\lambda$, where λ is the wave length of light, we find that any attempted measurement of the position of a ray by means of a device such as a narrow slit will be connected with a momentum exchange between the ray and the measuring agency, which is the greater the more accurate a position measurement is attempted. The analysis of experiments similar to the one just described reveals the fundamental significance of the uncertainty relations: We must accept the fact of the impossibility of any sharp separation between the behavior of optical objects (rays and waves) and the interaction with the measuring instruments which serve to define the conditions under which the phenomena appear.

What statements of physical significance can be made in our optical duality theory? At most we can define the probability P of finding a ray in some region of space and this likelihood is proportional to the square of the modulus of the wave function ϕ, i.e., $P \sim \phi^*\phi d^3x$.[13]

The operator relations (3) are invariant under unitary transformations and the transformed wave functions describe the same state of the optical system. Probability statements can also be made for the transformed wave functions. For example, the likelihood of finding a ray in a given momentum range is proportional to $\phi^{*\prime}\phi^\prime d^3p$ where $\phi^\prime$ is the suitably transformed wave function.

Assume that we are dealing with two rays of light, each associated with one or the other of two incoherent beams of light. In geometrical optics the trajectories of these two rays can only be described in a six dimensional configuration space or a twelve dimensional phase space. Since the beams do not interact, we may represent the total wave function of this system by the product of individual wave functions,

$$\phi = \phi_1(x_1)\,\phi_2(x_2)\,, \tag{6}$$

and the wave equation is now written in a multi-dimensional space,

$$(\nabla_1^2 + \nabla_2^2 + k_1^2\mu_1^2 + k_2^2\mu_2^2)\,\phi = 0\,. \tag{7}$$

If the wave function for the system is not written as the product of the individual wave functions for each light beam, the wave theory does not go over to the correct short wave length limit. In that limit, the ray dynamics must be described by a Hamiltonian which is the sum of Hamiltonians for each light beam taken separately,

$$H = H_1 + H_2 = \tfrac{1}{2}(p^2 - \mu^2) + \tfrac{1}{2}(p^2 - \mu^2)\,. \tag{8}$$

We also observe from (7), that if the two rays are identical ($k_1 = k_2$, $\mu_1 = \mu_2$), a new solution of the wave equation may be formed by inter-

151

change of coordinates, $\psi = \phi_1(x_2)\,\phi_2(x_1)$, and since the wave equation is linear, the sum or difference of the two solutions ϕ and ψ is also a solution. If we choose $\phi \pm \psi$ as the solution of the wave equation, we introduce an effective interaction (or coherence) between our rays. This coherence could never arise in geometrical optics; it has meaning only in our optical duality theory.

In the course of time, it was found that our optical duality theory could not explain all the experimental evidence. It was necessary to ascribe an intrinsic angular momentum (spin) to our rays which could not be given a classical (ray) interpretation. The spin is not a continuous variable as is the trajectory of a ray, but has only two values; one in the ray direction and a second opposite to that direction. This spin property of the ray is best represented by a matrix operator. As a further proof of the non-classical character of the spin, it is easy to show that its magnitude is proportional to $1/k$, the parameter which vanishes in the domain of validity of geometrical optics.

Our parable has come to an end, and we must now try to assess its meaning.

We have recreated the history of geometrical-wave optics and reinterpreted its mathematical symbolism so that both history and interpretation parallel the development of the quantum theory. Our re-interpretation has proved possible because we have examined only a truncated version of the classical Maxwell-Lorentz theory, a well-known deterministic theory; for knowledge of the electromagnetic field and all mass particle positions and momenta at one instant of time makes possible, in principle, the determination of the field and particle motions at all future times. And yet if we restrict ourselves to optics and continue to demand a complete description, in the Newtonian sense, of the rays of the electromagnetic field, we find that no determinacy is possible and that Bohr's complementarity description of the ray-wave duality is always a permissible consistent interpretation of the light ray experiments. However, rather than introduce the indeterminacy of the optical duality theory for the rays based on the dichotomy between the micro-objects and the measuring apparatus, it seems more plausible to consider the rays as secondary phenomena associated with the electromagnetic waves, which are to be considered as primary. The rays manifest themselves only in a certain limit of the wave motion, and the uncertainties associated with the ray description can be understood as reflecting a characteristic property of wave phenomena in general. Therefore, instead of a non-deterministic optical duality theory, we have the determinism of the wave theory (knowledge of the wave function ϕ). We also must resign ourselves to the fact that, once given the wave theory, we can never return to the determinism of geometrical optics, i.e. to a complete knowledge of the ray motions.

Can we extend this idea of wave determinism to the quantum theory of mass particles, and can we understand the particles as secondary phenomena associated with wave processes described by the Schrödinger wave function? The Heisenberg uncertainty principle would then be a consequence of the fact that the particles are simply the rays associated with a wave phenomenon. The distributions of particles that we observe would depend on the precise experimental situation, but the interaction between the measuring apparatus and the wave would determine what we observe, and this would be the only role assigned in the theory to the instruments of measurement.

Is such an interpretation of the quantum theory possible?

Before we answer this question, we must indicate a physical distinction between the rays of optics and the mass particles of the quantum theory. A mass particle always retains its identity.[14] On the other hand, in wave optics any tube of light can be turned into an infinite number of other tubes through successive diffraction and focusing. The concept of a light ray, therefore, does not completely correspond to that of a mass particle, for the concept of the ray exists only as a limiting process. It is only in the quantum theory of radiation that this continuous wave property of the field fails, since in the quantum theory the energy of the electro-magnetic wave appears only in discrete units.

In answer to our question posed above, we say that the quantum theory of a particle with mass corresponds to the short wave length limit of the classical electromagnetic field with the following subsidiary condition: The electromagnetic energy is incapable of continuous subdivision. Thus we may interpret the non-relativistic quantum theory as a deterministic wave theory, with the particles simply the rays associated with the wave phenomena, and the atomicity of matter assumed from the outset.[15]

If we adopt this view of the quantum theory, such questions as the "reduction of the wavepacket" no longer pose problems, for we have a deterministic wave theory with an assumed discreteness of the rays associated with the waves, but with knowledge of the ensemble of rays alone permitted by the theory. However, if we wish to go beyond the bounds of the present quantum theory, and explain the atomicity of matter, we need a theory whose imagery is no longer restricted to the primitive classical concepts of wave and particle.

We have completed a brief discussion of various interpretations of the quantum theory. Most of our attention was directed at unorthodox deterministic views of this theory. While we do not subscribe whole-heartedly to any one of these interpretations (we obviously favor the interpretation of the quantum theory as a complex classical field theory), we do feel that they are worthy of note because they direct attention to the use of physical images (models) in our attempts to analyze fundamental processes. Of course, such models are often utilized by adherents to the Copenhagen interpretation, but they are never taken seriously; in this interpretation only the mathematical scheme possesses true content.

For us, the major distinction between deterministic and orthodox views of elementary processes lies in the acceptance or rejection of the possibility of forming a mental image of such events. We are loathe to surrender our imagery on the basis of the success of the Copenhagen interpretation of the quantum theory, for, as we have shown, if we apply the arguments of the Copenhagen school to the description of optical phenomena, we can do away with the well-accepted wave model of these phenomena. It is possible to explain all the experimental data of optics by means of a duality interpretation and thus completely ignore the established classical concepts in this field. However, it is doubtful whether many physicists would find virtue in such an interpretation of classical optics. It is my feeling that perhaps we have too hastily deprived ourselves of the use of imagery on the level of quantum events. While the use of such imagery will hardly effect the status of the Schrödinger equation, it might very well provide us with new methods for finding future successful theories of elementary phenomena.

Stevens Institute of Technology, Hoboken, New Jersey

NOTES

1. V. Fock, *The Theory of Space Time and Gravitation*, Pergamon Press, New York, 1959, p. XVIII.
2. S. Körner, *Observation and Interpretation*, Butterworths Scientific Publications, London, 1957, p. 41.
3. P. A. Schilpp, *Albert Einstein: Philosopher-Scientist*, Library of Living Philosophers, Inc., Evanston, Ill., 1949, p. 3.
4. E. Schrödinger, *Brit. J. Phil Sci.* 3 (1952), 109, 233.
5. M. Planck, *Scientific Autobiography and Other Papers*, Philosophical Library, New York, 1949.
6. L. De Broglie, *Non-Linear Wave Mechanics*, Elsevier Publishing Co., Amsterdam, 1960.
7. M. von Laue, *Naturwissenschaften* 38 (1951) 60.
8. S. Körner, *Observation and Interpretation*, Butterworths Scientific Publications, London. 1957, p. 33.
9. This non-relativistic and static mechanical theory may be extended to the special theory of relativity and to true motion in time for the rays by modifying Fermat's principle to read' "the ray will take that path which accomplishes the motion in the least proper time".
10. The actual history of geometrical optics was quite different, since Hamilton's optical theory (embodied in eq. (1)) was formulated in the first half of the 19th century. However, the theory *might* have been created before Young's discovery of the wave-like nature of light.
11. At this stage of our argument, the rays are not to be confused with photons which only exist in a quantized theory. Our rays do not have all the properties of mass particles, a point which we shall amplify later.
12. $1/k$ plays the same role in optics as does Planck's constant h in the quantum theory.
13. The probability distribution of rays defined by $\phi^*\phi d^3x$ arises from the conservation law of electromagnetic field energy. We have chosen ϕ to be proportional to a single component of the electric field vector, so that $\phi^*\phi d^3x$ is proportional to the electromagnetic energy in the volume element d^3x. In the limit of geometrical optics, we relate the density of rays to the energy density of electromagnetic energy by arbitrarily assuming a certain number of rays to be associated with a given energy density of the beam. Hamilton's theory of geometrical optics which is embodied in eq. (1), also predicts an equation of continuity for rays. The conserved density of rays in the short wave length limit is roughly $\phi^*\phi$. The law states that the number of rays is conserved, and this conservation law is again a consequence of the conservation of electromagnetic energy.
14. In relativistic phenomena, particles may be created and destroyed, but our present discussion is reserved for non-relativistic effects.
15. Aspects of this thesis are in agreement with the views of Schrödinger; see note 4.

ARMAND SIEGEL

OPERATIONAL ASPECTS OF HIDDEN-VARIABLE
QUANTUM THEORIES
WITH A POSTSCRIPT ON THE IMPACT OF
RECENT SCIENTIFIC TRENDS ON ART

Presented March 27, 1962

The domination of the field of quantum physics by the "Copenhagen interpretation" [1,2] until recent years pushed the concept of the thing-in-itself, even that of objective reality, well into limbo as far as the outlook of theoretical physics was concerned. At first the wave function was even thought to be largely of subjective implication. When this misconception was finally cleared up, there still remained as an irreducible residue in this interpretation the idea that the wave function was strictly a record of experimental results. The objectivity that remained in quantum physics was thus of the nature of objective *appearance*, well removed from the inherent properties (if any) of the system under investigation.

Now the work of Bohm, [3, 4] and of Wiener and Siegel, [5, 6] has demonstrated by construction the possibility of theories which explain the experimental facts of quantum physics from a point of view consistent with strict determinism. Such theories are referred to as "hidden-variable" theories, this term referring to their most conceptual component. With such theories the concept of objective reality can make a tentative re-entry into quantum physics, and we shall here be concerned with the different ways in which this occurs in these two hidden-variable theories. I do not consider it necessary to defend the deterministic point of view, nor to discuss the frequent dogmatism of the Copenhagen school; both of these things have already been well and eloquently done by Feyerabend [2].

I shall first crudely sketch the two theories, with emphasis mainly on the loci of determinism in them. Bohm's theory, in its original form omitting spin (we do not need to consider the spin refinements), may be taken to start from the quantum-mechanical definitions of the mean

156

values of position $\vec{x}$ and momentum $\vec{p}$, for a "pure state" having wave function ψ:

$$\langle \vec{x} \rangle = \int \psi^* \, \vec{x} \, \psi \, \overrightarrow{dx} \tag{1}$$

$$\langle \vec{p} \rangle = \int \psi^* (- i\hbar \vec{\nabla}) \, \psi \, \overrightarrow{dx} , \tag{2}$$

where $\vec{\nabla}$ is the gradient operator, $\overrightarrow{dx}$ is the volume element, and the integration is over all space. The device that underlies Bohm's theory is to put both of these into the form of statistical averages of classical probabilistic form over an ensemble of particles distributed through space with a probability density

$$\rho(\vec{x}) = \psi^*(\vec{x}) \, \psi(\vec{x}) . \tag{3}$$

The average of a dynamical variable f can then be calculated if it can be put in the form of a function of position, i.e. $f = f(x)$; then

$$\langle f \rangle = \int \rho(\vec{x}) f(\vec{x}) \, \overrightarrow{dx} . \tag{4}$$

This implies that position alone is the hidden parameter in standard quantum mechanics. (It will be different in those hypothetical situations Bohm envisions, in which there are deviations from the distributions predicted by quantum mechanics.)

The mean value of position as given by equation (1) is already in the form (4), with

$$f(\vec{x}) = \vec{x} . \tag{5}$$

The mean value of momentum, as given by equation (2), is not explicitly in the desired form, but can readily be put into it: The imaginary part of the integral in (2) is zero, as is well known, so we can drop the imaginary part of the integrand; next we multiply and divide the integrand by ψ and write the result as

$$\langle \vec{p} \rangle = \int \psi^* \psi \, Re \left(- i\hbar \, \frac{\vec{\nabla}\psi}{\psi} \right) \overrightarrow{dx} , \tag{6}$$

which is of the form (4), with $f = \vec{p}$ and

$$\vec{p}(x) = Re \left(- i\hbar \, \frac{\vec{\nabla}\psi(\vec{x})}{\psi(\vec{x})} \right) . \tag{7}$$

Thus the "pure state" of quantum mechanics may be regarded as describing an ensemble of particles randomly distributed in space with the pro-

157

bability density $|\psi(\vec{x})|^2$. Each particle of the ensemble has a definite position $\vec{x}$ and a definite momentum determined by its position, $\vec{x}(\vec{x})$. Position and momentum mean values nonetheless agree with those postulated by quantum mechanics. Bohm then goes on to take up many other necessary considerations: How the dynamics of the particles of this ensemble can be stated in Hamiltonian form, adding to the potential $V(\vec{x})$ of the Schrödinger equation a characteristic "quantum potential"; how this classical ensemble theory can be reconciled with the quantum theory of measurement and with the disturbance usually postulated with the act of measurement, etc. I shall not discuss these matters, but merely emphasize the fact that the individual events of the ensemble in Bohm's theory are characterized in the terms of classical mechanics, and obey classical equations of motion. The random aspects of the theory do not change this, but merely put a probability measure on the events thus classically defined.

We now turn to the differential-space theory, proposed by Wiener and Siegel. It is like Bohm's theory in that the "pure state" of quantum mechanics is made to correspond to an ensemble of classical probability type, with probabilities or probability densities rather than probability amplitudes playing the fundamental role. But it differs strikingly in that the individual event is defined in terms of concepts taken from the mathematical theory of quantum mechanics rather than from classical mechanics. Since I have said that the differential-space theory differs basically from quantum mechanics, the last statement may sound confusing. We may put it this way: The differential-space theory borrows, selectively, from quantum mechanics for its elements. Some basic mathematical constituents are retained, as are some of the physical interpretations of them. The difference is that strict quantum mechanics (at least with pure states) makes statements about probabilities of events without defining a probability space, or sets of events, or individual events, within the probability space; while the differential-space theory does define a probability space with such sets. [1]

What the differential-space theory does take over from quantum mechanics is the use of the Hilbert space of wave functions (or state vectors), and (at least tentatively, in the present form of the theory, which is adjusted to give agreement with quantum mechanics) the representation of observables by linear Hermitian operators in this space. This is apart from the really only incidental (in principle, at least!) use of the wave function of the pure state in proving agreement with strict quantum mechanics; this the Bohm theory does too.

In the differential-space theory, physical observables are identified with random variables; we shall denote them by capital letters, e.g. R. The Hermitian operator which represents R in quantum mechanics will be denoted by the bold-face symbol R; and the eigenvalues of R, or the realized values of R, by the lower-case r (no confusion will arise from this double usage). Suppose the observable R to be measured in the ensemble represented by the state vector ψ, in the following precise sense: The measuring apparatus used is capable of resolving the value of R to within one of the sets $\Delta_1 r, \Delta_2 r, \ldots \Delta_m r$, but not better; as the standard interpretation would put it (and I invoke it at this point merely to utilize its familiar terms of reference in defining the kind of experiment I have in mind) the apparatus does not disturb the mutual phase relations, or coherence, of the probability amplitudes with respect to r, *within* these ranges, but it does destroy the coherence between any two *different* ranges. The sets given are disjoint, and assumed to exhaust the entire eigenvalue spectrum of R.

The Born statistical postulate of quantum mechanics then states that the probability that R lies within an arbitrary set $\Delta_k r$ is

$$\text{Prob. } (r \in \Delta_k r) = \| P(\Delta_k r) \, \psi \|^2 , \tag{8}$$

where $\| f \|^2 = (f, f)$ stands for the squared norm of f and $P(S)$ for the projection operator onto the manifold in Hilbert space corresponding to the eigenvalues r belonging to the set S. For example, in the well-known case in which $r = x = $ a position variable, $\psi(\vec{x})$ is the ordinary wave function and

$$\| P(\Delta_k \vec{x}) \, \psi \|^2 = \int\limits_{\Delta_k \vec{x}} | \psi(\vec{x}) |^2 \, \overrightarrow{dx} \tag{9}$$

the subscript to the integral sign indicating the region of integration.

Equation (8) constitutes a necessary condition of the differential-space theory, in that it must be possible to set up distributions in the probability space postulated by theory, which reproduce the statistical properties stated by equation (8) for all observables. I shall not give mathematical details about how the probability space is set up, since these are given in the references cited; by restricting the discussion to *what* is done, rather than *how*, I hope to show with a minimum of distracting technicalities how the theory depicts the physical world. Suffice it to say that the probability space is an extended form of the Hilbert space of wave functions, with a certain real probability measure defined on it; with the probability measure incorporated, the space is called "differential space". It is possible with this measure to compute the probability of a very wide variety of subsets of differential space.

Let us represent points in differential space by the symbol α. This represents an individual system, having definite values of any desired observable; but only in a rather un-physical sense, in that α is defined *a priori* only in terms of the coordinates of the point with respect to a complete set of axes of Hilbert space, and it does not appear that these have any simple meaning in terms of observables. I think that α as thus directly given must be considered to be of purely formal import. The physical content of the theory enters through an algorithm (the "poly-chotomic algorithm" of the references cited; the dichotomic algorithm may also be used, but then the eigenvalues of R may only be divided into two sets) by which a realized value of the observable being measured is assigned to α. We must at this point make an important qualification of the meaning of the last remark. In terms of the measurement as we have defined it above, the realizable values of the observable are the sets $\Delta_1 r$, $\Delta_2 r, \ldots \Delta_m r$. That is, the polychotomic algorithm assigns one of these *sets* to the point α, not a unique value of R (unless the set should happen to consist only of a single value of R). This mapping of sets of eigenvalues into points of differential space thus defines a set-valued random function on this space. Such a mapping exists for any R having a soluble eigen-value problem, and for any partition of the eigenvalue spectrum of R into a finite number of measurable sets. The aggregate of all set-values, of all observables, assigned to α obviously constitutes complete information about the outcomes of all possible measurements on the system represent-ed by α, with no restrictions on simultaneous measurability of variables. These set-valued random variables, so closely related to the observables of standard quantum mechanics, are hidden variables of differential-space quantum theory.

The basic theorem of differential-space quantum theory, in the poly-

160

chotomic form, states that the probability measures of the sets of differential space whose points are assigned to a set $\Delta_k r$ agree with equation (8). The mapping of eigenvalue sets into points α of course depends, parametrically, on the assumed wave function. We are here speaking of ensembles designed to agree, by construction, with the distributions defined by quantum mechanics. In principle, as in Bohm's theory, one is free to imagine many ways in which distributions incompatible with quantum mechanics could be set up.

The way in which determinism comes into this scheme is almost trivial. Let the value of the time at the present instant be zero... i.e., take the present moment as the origin of the time axis. It then turns out that the set-value, for a given partition of eigenvalues, of R associated by the polychotomic algorithm to α at time t (which may be earlier or later; the theory is perfectly symmetric to time-reversal) is just the *present* (i.e. at $t = 0$) set-value of the operator

$$R_t = e^{-\frac{t}{\hbar}H_t} \, R \, e^{\frac{t}{\hbar}H_t}$$

provided the eigenvalues of R_t are partitioned into sets in the same way as those of R (the eigenvalue spectra of the two operators are necessarily the same). R_t happens to be the operator into which R develops after time t, starting at zero time, in the Heisenberg picture of standard quantum mechanics; but in the scheme we are describing it is to be taken at zero time, and the quantity t which it contains is to be regarded as playing a parametric role.

The determinism so demonstrated is, considered purely as determinism, somewhat disappointing. The future value of R does not develop out of its present value. It is instead built into the present situation in fully developed form. To predict a number of future observables, it takes exactly the same number of present observables, and each of these applies only to one future observable, at one future time. And incidentally, it may be noted that this scheme leans rather heavily on the notion that every hypermaximal Hermitian operator (i.e., Hermitian operator having a soluble eigenvalue problem) possesses a corresponding observable; which is a rather extreme exploitation of the correspondence between idealized experiments and mathematical formalism that was introduced into physics by quantum-mechanical theory.

I think the disappointing "built-in-ness" of this result comes only from

161

pushing the deterministic picture too far. It is satisfying to know that the ensemble in the differential-space theory is built up out of sharp individual cases. But the individual case, perfectly predicted only by an impossible proliferation of data, cannot in reality be isolated. A subensemble of systems, somewhat spread out in the probability space, behaves more reasonably. Although the present value of an observable for the individual case does not *rigidly* determine its future value, there is a positive correlation between these values, more or less diminishing with time. Using subensembles, advantage could be taken of this correlation to formulate unforced, un-artificial theories of prediction. It seems in the spirit of the probabilistic physics of the future that (as Wiener has pointed out) the imperfectly defined ensemble of systems is a more natural entity to work with than the overidealized, perfectly defined individual system.

One further observation: With the advantage we have taken of a hypothetical simultaneous definability of any number of observables, it is really not surprising that ensembles consisting of sharply determined individual cases, yet agreeing statistically with quantum mechanics, can be set up. For imagine a "pure state" at $t = 0$ with wave function ψ, and a set of measurements at $t = 0$ and at $t = T$ to be made on systems of the ensemble, hypothetically without mutual interference. Set up a probability space with, as random variables, the observables to be measured at $t = 0$ and the transforms into "present" operators (à la equation (10)) of those to be measured in the future (i.e., at $t = T$). Then, by simple *fiat*, postulate distributions for these observables which agree with the Born postulate. This defines an ensemble agreeing with quantum mechanics and yet having sharp values [2] of all present observables. To make it deterministic also, at least with respect to observables to be measured in the future, simply provide that the future value of an observable R shall be defined as the present value of the observable R_T. By extending this cumbersome but essentially trivial procedure to all operators and all times, which of course requires disregarding difficulties due to the fact that the number of times and the number of operators involved has the power of the continuum (or even higher in the case of the operators?), one might imagine a complete, probabilistic but "crypto-deterministic" (the term is due to Whittaker, see [6]) description to be achieved. This is, of course, an *ad hoc* theory to the ultimate degree, infinitely worse than the Ptolemaic solar system. But despite its crudity it does heuristically indicate, without any deep mathematics, why there is no absolute logical bar to a hidden-variable theory of quantum phenomena. Von Neumann's celebrated "disproof" of such a theory rested on a set of assumptions so restrictive as to be virtually equivalent to quantum mechanics; in particular, the hypothesis of von Neumann most strikingly disobeyed by this *ad hoc* theory and by the differential-space theory is that sums of observables are necessarily associated with the sums of their associated operators.

Lest the faults of the above *ad hoc* theory be attributed to the differential-space theory, I hasten to point out how they differ. While the former requires a new rule for modifying the ensemble each time a new observable is introduced as such (an arbitrary decision must then be made as to the correlation of the new observable with those already incorporated), the differential-space theory satisfies the basic requirement of an acceptable scientific theory that every detail of such a theory be derivable from a small number of simple postulates. Of course, the differential-space theory thereby contains the correlations of observables as an already built-in feature. How great a range of variation there may be for the form of these correlations, consistently with the general scheme of a hidden-variable theory, is an interesting question to which I do not know the answer.

At this point I want to shift the emphasis from the substance and particular features of the differential-space theory, to an examination of it simply as a concrete exemplar of certain characteristics which embody interesting trends in scientific thought. By stating my purpose in this way, I hope to avoid the appearance of advocating *this* theory. Such advocacy would certainly be premature both because the differential-space theory as such is in a very preliminary state, with many apparently awkward features, and because there is as yet no discernible trend in empirical physics calling for explanation in terms of hidden parameters.

Right at the start of this more general discussion, I think I can state its basic thesis quite simply: The differential-space theory utilizes certain positive aspects of operationalism [3], while rejecting other, negative, features of it which were incorporated into quantum mechanics. [4] Where the differential-space theory utilizes operationalism positively is in the identification of the result of the act of measurement, *precisely specified*, with the elements used to describe the physical situation. In this it is very much the child of quantum mechanics. As in quantum mechanics, "I am measuring R" is not sufficiently precise for the theory. Consistent with the statement are an infinity of ways of partitioning the eigenspectrum of R, such that only within any one subset of the partition does the measuring instrument (I will here use the language of quantum mechanics, which has a satisfactory operational meaning for systems which are correctly described by it) preserve the previously existing phase relationships among the probability amplitudes of the various values of R. Both theories distinguish these various operational procedures quite carefully, and assign different experimental outcomes to them.

By accepting all possible measurements as hypothetically performable without mutual disturbance, the differential-space theory leads to an even more striking possible consequence of this aspect of operationalism than is found in quantum mechanics: Consider two such measurements, one in which a certain interval ΔR forms one of the basic (not further divided) elements of the partition, and another, in which the interval ΔR is not basic, but is divided into (say) two basic subintervals, $\Delta_1 R$ and $\Delta_2 R$. In the differential-space theory, an individual system belonging to ΔR for the first measurement [5] need not belong either to $\Delta_1 R$ or $\Delta_2 R$ for the second; and vice versa!

This result may seem absurd to you, and both Professor Wiener and I have felt real difficulty in accepting it. Nevertheless I can see no reason to rule out this possibility. And there is always the possibility that we have here one of those situations in which a rigorous and complete mathematical formalization consistent with the experimental facts leads to previously hidden experimental consequences; although the possibility of an experimental verification of such an effect does at the moment seem quite remote.

This effect cannot, of course, show up in standard quantum mechanics because either measurement is considered to change the wave function of the system in such a way as to prevent the anomalous result in the other. But such a change in the wave function represents a physical disturbance of individual systems of the ensemble from the hidden-parameter point of view, changing the properties of systems which would display anomalous behavior. The probabilities, of course, are free of any anomalous properties: Since there is always agreement with quantum mechanics (as long as the polychotomic or dichotomic algorithms are used), the probabilities of $\Delta_1 R$ and $\Delta_2 R$ in the second measurement add up to just the probability of ΔR in the first.

As long as the "conditions which define the possible types of predictions regarding the future behavior of the system . . . constitute an inherent element of the description of any phenomenon to which the term 'physical reality' can be properly attached" [7], which I take to be a fair statement of the operational point of view in quantum physics, I do not think the above-described effect in the differential-space theory constitutes an antinomy. The "conditions" in the two experiments are different, (they are different even in the experiments as they would be performed today, since they produce different types of incoherence relations among probability amplitudes), and if we then admit that they constitute different "inherent elements", we cannot *a priori* require any preconceived forms of consistency between them. This is a far cry, indeed, from classical mechanics. But note that, at least as far as the form of differential-space theory tailored to agree with quantum mechanics goes, there is a good probabilistic correlation between the system's belonging to ΔR in the first experiment and belonging to at least one of the constituent subintervals of ΔR in the second; which makes the result easier to accept on grounds of physical intuition.

The negative feature of operationalism that is rejected in the point of view here expressed is the assertion, incorporated into quantum mechanics, that concepts which are operationally undefinable in terms of *present-day* experimental techniques are absolutely to be ruled out of any theoretical description, even if used in a logically consistent way. The measurability of "non-commuting" observables without mutual disturbance is such a concept. But – to reiterate the usual arguments – it is conceivable that radically new experimental techniques may emerge in the future, which would extend the possible operational definitions of observables so as to make possible under certain conditions a weakening of the hypothetical mutual disturbance of measurements. In other words, what is un-operational today may become operational tomorrow.

By so formulating the situation, we are on the verge of saying that this "negative" feature of operationalism, the exclusion of currently impossible measurements, is not really part of operationalism. But this is done by introducing a distinction between "hypothetically" operational and "actually" operational which is foreign to the spirit of the doctrine, at

165

least as it is usually accepted; the usual formulation would, I think, insist that the trend of the more recent and highly developed experimental techniques establishes a canon of operational acceptability which excludes concepts which appear to reverse this trend.

I do not wish to re-define operationalism (at least not to this extent), and therefore shall not insist that the differential-space ideas are strictly operational throughout. But if theories are to go beyond present facts at all, they must adopt some hypotheses which will affect even the realm of operational concepts. However clear the trend of the future may seem to be, the history of science shows amply how the dogma of one period is often rejected in the next. One can agree whole-heartedly that the splendid achievements of quantum mechanics as now formulated express a *Zeitgeist* in operationalism quite foreign to the hidden-variable idea. But the achievements of thermodynamics in the nineteenth century also had a compelling intellectual force, and Mach read them as pointing away from statistical mechanics:

"The mechanical conception of the Second Law through the distinction between *ordered* and *unordered* motion, through the establishment of a parallel between the increase of entropy and the increase of disordered motion at the expense of ordered, seems quite artificial. If one realizes that a real analogy of the entropy *increase* in a purely mechanical system consisting of absolutely elastic atoms does not exist, one can hardly help thinking that a violation of the Second Law – and without the help of any demon – would have to be possible if such a mechanical system were the real seat of thermal processes. I agree here with F. Wald completely, when he says, 'In my opinion the roots of this (entropy) law lie much deeper, and if success were achieved in bringing about agreement between the molecular hypothesis and the entropy law, this would be fortunate for the hypothesis, but not for the entropy law'." [8]

If I may call Mach's point of view operational, we may say that developments which followed his above-quoted remark, most strikingly the discovery of the Brownian motion and Einstein's incorporation of it into statistical mechanics, gave an operational meaning to the "hidden, invisible motions" *(unsichtbare verborgene bewegungen)* [9] which were not operational when Mach wrote *Theorie der Wärme*. This certainly reversed

166

the trend which thermodynamics seemed to embody. Yet thermodynamics and statistical mechanics have since learned to live together, and to thrive in so doing.

With this supporting predecent, let me sketch a hypothetical reconciliation of the hidden-parameter idea with operationalism. Suppose we take the latter to be summed up in the simple statement, "The concept is synonymous with the corresponding set of operations as pictured by the mind".[6] There is nothing, at least in this statement, which forbids the conjecturing of as yet unknown operations. Now, if indeed it is in the nature of physical systems that certain measurements are incompatible with one another, such a fact could be expressed in a theory in which the possible and the impossible situations – both operationally described – are represented as *conceivable*, and the rejection of the impossible ones appears as a choice of nature rather than as a logical impossibility. Classical statistical mechanics, and Boltzmann's H theorem in its finally satisfactory form, did just this with respect to the "impossible" decrease of entropy; and the violations of the second law of thermodynamics turned out to be not impossible but merely improbable (and unexploitable). A similar outcome is what advocates of hidden-parameter theories are betting on in quantum theory.

Now I want to return to the question of the "thing-in-itself" and objective reality, raised in the first paragraph. I think that the hardest thing about the Copenhagen interpretation for many intuitive, or independent-minded, or (if you will) obstinately soft-headed physicists to take, was the virtual denial of the thing-in-itself; the assertion that nature herself denied us even the right to think about objects in themselves, if we wanted to think scientifically. Not only was the physical system defined in terms of measurements, but it was so constituted that not even in thought could it ever be boxed in to anything more precise that a certain minimum squiggly thing. I think this is the grievance that Einstein, Podolsky and Rosen [11] brought to precise form in their claim that quantum mechanics was not a "complete" theory.

Let me now violate my own announced principles, and invoke a *Zeitgeist* which seems to favor the form (if not substance) of the differential-space theory over Bohm's type of theory. When the thing-in-itself is a classical point particle, its denial in the context of modern quantum physics, and operational developments in general, certainly seems to

167

have some justification. The fantastic developments of recent times in experimental precision and refinement seem to leave no room for such a baroque entity as a point particle in an elaborate potential field. Even granting the surprisingly successful transferral of the potential field and Maxwell's equations to quantum mechanics in the twenties and thirties, the logic of modern developments seems to point to more non-commital concepts, as illustrated by the S-matrix and dispersion relations.

I therefore think it an interesting, if limited, achievement of the differential-space theory that it reconciles operationalism with the possibility of independently-existing objects of arbitrarily precise properties; that it may give the "thing-in-itself" respectability within the "logic of modern physics". Note that the phrase used is "arbitrarily precise", rather than "absolutely precise". This is because the number of independent observables is infinite, and a hypothetical sequence of measurements of successive observables, even if arbitrarily sharp (which in itself requires an infinite limiting process for a *single* operator), will not converge to a completely defined system. I think this is a satisfying feature, in that it allows within the theory itself for the possibility that no amount of measurement may exhaust the properties of objective reality.

In taking the picture sketched above to provide a satisfactory common ground between operationalism, and the incorporation of objective reality into a scientific theory, I assume that the role of the thing-in-itself is secure if there is no bound to the degree of its definability, even if it is not definable in finite terms. Where hidden-variable theories, in general, constructively demonstrate the lack of any necessary logical connection between quantum phenomena and indeterminacy, the differential-space theory further shows that there is no necessary conflict between an operational description of quantum phenomena and the explicit representation of objective reality.

POSTSCRIPT ON THE IMPACT OF RECENT SCIENTIFIC TRENDS ON ART

The choice of an operational pathway toward the scientific description of realities, which the above essay suggests may be open to physics, stimulates thoughts on the esthetic status, if any, of future science. Where esthetic considerations are brought up in scientific discussions, the tendency is (in my experience) to extol the beauty and harmony of the theory under discussion. If the harmonization of superficially conflicting facts is an esthetic achievement in the broad sense, any successful scientific theory is an esthetic achievement. But in an age where scientific thought is as highly developed as it is in ours, and permeates (in however vulgarized form) so many other areas of thought, the esthetic implications of a scientific theory are not to be so globally assessed.

The truly important esthetic impact of science lies not in the satisfactions it gives to scientists, or even to non-scientists vicariously participating in science, but in the way in which it shapes the various world-views that are currently tenable in a culture. This is a point of view which scientists seldom recognize, but which is well appreciated elsewhere; as for example, by more than one critic of the "New School" of literary criticism [12].

The highly developed science of what we think of as the scientific era has always stood in a state of conflict with the artistic spirit. I take this as a starting point, and it does not contradict this basic view to recognize that many scientists are artistically sensitive persons who would moreover like to regard science and art as different ends of a cultural continuum, or that many artists have exploited their sense of an antagonism between science and art for artistically creative purposes.

One way of stating this conflict is to say that in the view of the artist, science incessantly encroaches, both materially and spiritually, on the areas of undefined and uncategorized pure non-rational feeling which are the province of art. I. A. Richards, in *Science and Poetry* [13] outlined this problem very clearly two decades ago. In fact, I would like here to quote, in support of the reasonableness of my own fears for the future of the arts, one sentence from this book: "It is a possibility to be seriously considered that Poetry may pass away with the Magical View of the universe." Richards' proposed solution, that the worlds of science and poetry are really distinct if only they are correctly understood, seems to me the most promising one. But its implementation presents immense difficulties. Many of these are pointed out by Richards; and my aim in this postcript is to call attention to further problems raised by the changing nature of science and the world it is creating.

Since Richards wrote his book, it has become even more difficult to feel truly confident that the worlds of art and science do not overlap. As he feared, science has so much extended its scope that the artist must worry much more than before about straying into the wrong territory, or about the credence he can expect from his audience for the "truth" of the pseudo-statements [7] he makes.

Richards' argument for the *a priori* separateness of art and science was mainly based on the difference in the kind of statements each makes about the world. Another source of hope for art in this (for it) difficult contest might seem to lie in the insights into the limitations of science that have been gained in recent decades, which arise from increasing tendencies toward the axiomatization of science. That is, once one becomes aware of the dependence of science on controllable situations, of its ingrained habit of treating the real world on an "as if" basis, so that the grasp of its theories is just equal to the extent to which the chaotic real world coincides with its neat, finite pictures, it may seem quite unnecessary for art to feel that it is being crowded out of the world by science. But, like any purely rational argument, such a point of view does not provide art with the *emotional* support which it needs. In fact, it may work quite the other way: the careful self-limitation of modern science is not a weakness, but really one of the secrets of its success. And, even disregarding envy (a motive often attributed to artistic critics of science), the success of science creates a discouraging atmosphere for art precisely when this success arises from a drastic axiomatic simplification of a world which art prefers to regard as infinitely complicated and mysterious.

Art moreover finds it unbearable to compete with technology (which I here take to include science). I derive this observation from the fact that the graphic arts have abandoned realistic portraiture and landscape representation, as a major area, to photography. I think this retreat was an unnecessary one, in that the achievements of classical art in these areas have never been equalled by photography. The fact that it was made reveals the defensiveness of art in the face of technology.

So it is with the field of human behavior. As psychology, psychoanalysis and cybernetics increasingly dominate the field of the description of overt human behavior, the *avant garde* of art retreats (advances?) into the field of the "absurd', where for the time being science cannot follow. [8] And this happens notwithstanding the truths which still elude these scientific fields, and which may always continue to do so.

Art is affected by science through the *scope*, the *world view*, and the

170

success of the latter. Richards has discussed the question of scope, and argued (quite reasonably) that there is little overlapping of scope between them; but I have here presented the counterargument that the *success* of science reverberates so intensely through the world that the influence of science on art is far greater than its limited scope, as defined by considerations of intellectual subject matter, would warrant.

The impact of the success of science, or what amounts to the same thing for purposes of this discussion, its *intensity* as a functioning social institution, may also help to resolve the following paradox: That art seemed less alienated by the mechanistic, literal theories of earlier eras of science than it is by the seemingly more cautious operational theories of the current era. Although a man was represented as a sort of elaborate clockwork by the science of the nineteenth century, this was not so disturbing, in view of the extent to which this earlier science fell short of precisely implementing this picture, as the modern description, which is in less literal terms but is far more advanced in its empirical progress.

Moreover, to the extent to which art does respond to the literal content of science, it does not do so with respect to its face value. The strict and detailed mechanism of classical physics, which claimed to reduce the behavior of the entire universe to the precise motions of atoms in potential fields, might seem more depressing to a believer in the unique and irreducible qualities of the human soul than the seemingly less concrete theories of today. It seemed to leave no room at all for the soul, and traditionalists of art and religion did not fail to observe this and condemn it. Have modern theories, with their greater sophistication and less sweeping claims, led to a renewal of confidence on the part of artists? I do not sense any such change. Compared to the modern theories, the older ones had at least the charm of a metaphor. If man was a mechanism, at least he was one on the model of the stars and the planets. Today, on the other hand, the planets seem reduced to collections of data fit to be fed to a computing machine.

The trouble is, perhaps, that while the modern style in science has none of the picturesqueness of the old one, scientific theories have *not* ceased to be metaphors in the world as art sees it. Or as we all see it: anyone who has seen a bright child pretend to be a robot ("Charge my batteries; turn me to the right...") may know what I mean.

A theory of reality which defines it solely by measuring instruments, however elegant its mathematical formulation, seems from the point of view of art a ramshackle vision. Our metaphors are bound tighter and tighter to the practical world by science. The world of affect, which lies outside the logical realm of applicability of science, nonetheless resonates powerfully to its non-logical implications. Any view of the social responsibility of science will have to take this into account.

Boston University, Boston, Massachusetts

NOTES

1. It does not define individual events, as we shall see, in the traditional way, but through a "limit concept".
2. By this we mean not infinitely sharp, but mutually sharper than permitted by the uncertainty principle.
3. Professor A. Grünbaum has drawn my attention to the fact that my usage of the word "operationalism", while agreeing with that of most physicists, is quite different from that accepted by philosophers of science. The latter define operationalism as the doctrine that the literal operations themselves must generate all theories, with the guidance of the mind but without the intervention of mental constructs containing any elements foreign to the literal operations themselves. To me, and I think to most physicists, operationalism means the construction of theories out of concepts which correspond, as well as may be possible, to the literal operations. The autonomous elements of the theory are the concepts, not the operations; and the concepts are, so to speak, only pictorial representations of the operations.
4. The words "positive" and "negative" are not intended to convey value judgments here.
5. I purposely avoid saying "an individual system for which R belongs to $\triangle R$ for the first measurement", since it is just the point of this sentence that the first and the second measurement only assign the system to intervals of R, not to individual values of R within these intervals.
6. This differs from Bridgman's definition [10] of operationalism by the addition of the words "as pictured by the mind". These are inserted for consistency with the point of view stated in the footnote on p. 12.
7. Richards' term for the kinds of statements characteristic of poetry as distinguished from science.
8. Richards feared that some of the arts might destroy themselves by challenging science on the ground where it was supreme. I am concerned here with the opposite – namely that they retreat, rightly or wrongly, further than "pure reason" would seem to require.

REFERENCES

[1]. Essays by N. Bohr (pp. 201–241) and A. Einstein (pp. 665–688) in *Albert Einstein, Philosopher-Scientist*, (Ed. by Paul Arthur Schilpp), The Library of Living Philosophers, Evanston, Ill., 1949; and references therein cited.

172

[2]. P. K. Feyerabend, Niels Bohr's Interpretation of the Quantum Theory, in *Current Problems in the Philosophy of Science*, New York, Henry Holt, 1961; *Problems of Microphysics*, in Pittsburgh Publications in the Philosophy of Science, Vol. 1, Univ. of Pittsburgh Press (to be published).

[3]. D. Bohm, A Suggested Interpretation of Quantum Theory in Terms of "Hidden Variables", Parts I and II, Physical Review 85 (1952) 166–193.

[4]. D. Bohm, *Causality and Chance in Modern Physics*, Princeton, Van Nostrand, 1957.

[5]. N. Wiener and A. Siegel, A New Form for the Statistical Postulate of Quantum Mechanics, *Physical Review* 91 (1953) 1551.

[6]. N. Wiener and A. Siegel, The Differential-Space Theory of Quantum Systems, *Nuovo Cimento*, Ser. X, Vol. 2 (1955) Supplement 4..

[7]. N. Bohr in Schilpp collection (Ref. 1).

[8]. Ernst Mach, *Principien der Wärmelehre*, Leipzig, 1896, p. 364.

[9]. Ernst Mach, *loc. cit.* p. 363.

[10]. P. W. Bridgman, *The Logic of Modern Physics*, New York, Macmillan, 1927.

[11]. A. Einstein, B. Podolsky and N. Rosen, Can the Quantum-Mechanical Description of Reality Be Considered Complete? *Physical Review* 47 (1935) 777.

[12]. Stanley Edgar Hyman, *The Armed Vision*, Knopf, 1948.

[13]. I. A. Richards, *Science and Poetry*, Kegan Paul, 1935.

173

COMMENTS

ABNER SHIMONY

It is usually stated that the Copenhagen interpretation of quantum mechanics is the orthodox interpretation, held by almost all physicists. The statistics of our colloquium make one wonder about this: Two speakers and a commentator, chosen at random from the ensemble, reject the Copenhagen interpretation, and one begins to ask where any of the orthodox are to be found.

Before commenting on the specific proposals of the speakers, I should like to review briefly two features of the Copenhagen interpretation which seem disturbing to me, for such a review will explain my sympathy for alternative interpretations.

(1) The first disturbing feature of the Copenhagen interpretation is the sweeping claim which it makes about the limits of possible experiments. These claims are based on two very firm premisses: (a) the indivisibility of quanta in all interactions, which of course includes the interactions of microscopic systems with measuring instruments, and (b) the dismissal of any theoretical attributions to physical reality which are in principle incapable of experimental detection. (The second premiss does not presuppose an operationalist point of view; it can also be founded on the conviction that nature does not consist of insulated layers, some of which are completely unaffected by some of the others.) From these premisses all the exponents of the Copenhagen interpretation conclude that there can be no "hidden" characteristics of an individual system which determine the outcomes of all possible measurements; the outcomes of some, in fact most, measurements are irreducibly chance matters. I find this conclusion unconvincing primarily for a negative reason: that nothing but schematic proofs have ever been provided to show that the existence of indivisible quanta prevents the detection of such hidden characteristics, regardless of the subtlety of the apparatus. To be sure, there is Heisenberg's gamma-ray microscope, and Bohr has devised some ingenious thought-experiments. But nothing like a comprehensive argument exists, comparable to the impossibility arguments regarding ruler-and-compass constructions or

174

even (to use a fairer comparison) to arguments for the non-existence of Maxwell demons.

(2) According to the Copenhagen interpretation it is apparently necessary to refer to the perceptions of human observers in order to give a complete formulation of quantum theory. Without the ultimate act of observation the evolution of a composite system (including object and apparatus) is governed by the Schroedinger equation, and hence all terms in an initial superposition of eigenstates are propagated. Only the ultimate act of observation selects one of these, thus "reducing the wave packet". One can't help asking naive questions such as: Did no reductions of the wave packet occur before human beings evolved? Did the result of the measurement depend upon an examination of the photographic plate which was supposed to register the result? To these naive questions sophisticated answers have been given, to the effect that the ultimate act of observation makes little difference in practice to the physical system, because the intervention of macroscopic apparatus has for all practical purposes destroyed the coherence of the superposed eigenstates. I am not content with these answers, partly because "in practice" implies an attitude towards the activity of science and the objects of scientific inquiry which I reject.

The attractiveness of hidden variable theories, in my opinion, lies not so much in their reinstatement of determinism as in their escape from these two disturbing aspects of the Copenhagen interpretation. Hidden variable theories of all varieties are critical of dogmatism regarding the limits of possible experiments, and Bohm in particular has tried to indicate at what points the detection of hidden variables might be possible. Furthermore, the hidden variable theories are generally free from the intrusion of subjective elements into the formulation of the physical theory.

Nevertheless, the hidden variable theories exhibit a characteristic vulnerability. The Copenhagen interpretation provides a systematic scheme for rejecting a certain class of questions as in principle unanswerable, whereas a hidden variable theory assumes an obligation to answer them – or at least, in its tentative form, to sketch an answer. Most of the questions which I shall put to Profs. Schiller and Siegel are really requests for plausible outlines of explanations which the orthodox theory does not have to provide.

The differential-space theory seems to me remarkably successful as a formal method of attributing deterministic behavior to individual systems, at the same time making statements about probabilities and expectation values of observables which completely accord with quantum mechanics. In particular, it agrees with those probability evaluations in which coherence effects are important – i.e., in which probability amplitudes rather than probabilities must be added. But one can't help wondering *why* these effects occur. Why should the ensembles be so arranged that the two-slit pattern should exhibit interferences? One would like an answer either in dynamical terms – like Bohm's quantum potential – or in statistical terms, or in some combination of the two. Perhaps even more puzzling, from the point of view of a hidden variable theory, is the great success of quantum mechanics in assuming only symmetrical or antisymmetrical states when dealing with systems of identical particles. This success is plausible in the orthodox interpretation because no particle has a completely determined trajectory, and hence it makes sense to say that a particle loses its self-identity. But in a hidden variable theory the particles maintain their self-identity, and some additional explanation – perhaps some mixing mechanism – seems required in order to explain symmetrization and antisymmetrization.

One also hopes for some further explanation in the theory which Prof. Schiller has sketched. It is perhaps not quite correct to characterize his proposal as a hidden variable theory, for it is rather a theory in which the "wave" or "Schrödinger field", governed by the Schrödinger equation, is conceived as real in the same classical sense that the electromagnetic field is real. This is very similar to Schrödinger's original interpretation of the wave function, which was abandoned because wave packets – which Schrödinger took to represent particles – always dispersed, whereas physical particles do not disperse. The imposition of a subsidiary condition that energy is carried in quanta leaves open the question of the exact relation of the wave to the quantum: why is it that the wave passes through both slits but the quantum through only one? It seems hard to stop at the point which Prof. Schiller has chosen, for one seems impelled either to go over to the Copenhagen interpretation, or else to introduce variables which are not entirely represented by the "Schrödinger field".

I should like to return to Prof. Siegel's paper with a remark of a different kind. The precise content of his statements concerning measurements of an observable R to different degrees of refinement was not clear to me. The two operators

$$\sum r_i P_{(\Delta r_i)} \quad \text{and} \quad \sum r_j P_{(\Delta r_j)}$$
$$\Delta r_i = (r_i, r_{i+1}) \quad \Delta r_j = (r_j, r_{j+1})$$

(where the second subdivision is a refinement of the first) are commuting operators, though the latter has eigenvalues which the former does not have. Hence they are simultaneously measurable, according to the usual formalism of quantum mechanics. Since the differential-space theory is supposed to agree with ordinary quantum mechanics whenever the two theories overlap, I am puzzled by the statement that if the system is found to be in Δr_1 and if $\Delta r = \Delta r_1 \cup \Delta r_2$, then it need not be in Δr.

Finally, after all these essentially conservative comments, I should like to be allowed one radical remark. It seems to me that the hidden variables proposed by Bohm, De Broglie, Siegel, and Wiener are perhaps much too similar to the classical quantities of pre-quantum physics and of the present quantum physics. As a result, critics of the hidden variable theories are able to raise the question: what phenomena do you propose to explain? And the answer is usually that there will be some phenomena to explain when bigger accelerators are built, etc. However, there are gross phenomena which present theories of nature are not remotely capable of explaining, and one doesn't have to look far for them – namely, feeling and thought. Is it not possible that the hidden variables to be sought, or at least some of them, are of a "psychological" character, so that the ultimate entities might resemble the monads of Leibniz or the occasions of Whitehead? Eventually, I believe that a "unified theory" embracing physics and psychology will be found, and the psychological element may be preponderant. It is, of course, a quite different supposition that such a theory is precisely what is required in order to solve the dilemmas which now confront us in quantum mechanics, and it is very likely that several levels of physical theory separate contemporary physics from such an ultimate "unified theory".

Massachusetts Institute of Technology, Cambridge, Massachusetts

ADOLF GRÜNBAUM

THE FALSIFIABILITY OF THEORIES:
TOTAL OR PARTIAL?
A CONTEMPORARY EVALUATION OF THE
DUHEM-QUINE THESIS *

Presented April 26, 1962

It has been maintained that there is an important *asymmetry* between the *verification* and the *refutation* of a theory in empirical science. Refutation has been said to be conclusive or decisive while verification was claimed to be irremediably inconclusive in the following sense: If a theory T_1 entails observational consequences O, then the *truth* of T_1 does *not*, of course, follow *deductively* from the truth of the conjunction

$$(T_1 \rightarrow O) \cdot O$$

On the other hand, the *falsity* of T_1 is indeed *deductively inferable* by *modus tollens* from the truth of the conjunction

$$(T_1 \rightarrow O) \cdot \sim O.$$

Thus, F. S. C. Northrop writes [1]: "We find ourselves, therefore, in this somewhat shocking situation: the method which natural science uses to check the postulationally prescribed theories... is absolutely trustworthy when the proposed theory is not confirmed and logically inconclusive when the theory is experimentally confirmed."

Under the influence of the physicist, philosopher of science and historian of science Pierre Duhem[2], this thesis of asymmetry of conclusiveness between verification and refutation has been strongly denied as follows: If "T_1" denotes the kind of individual or *isolated* hypothesis H

---

* Portions of the present essay are drawn from earlier publications by the author as follows: A. Grünbaum, The Duhemian Argument, *Philosophy of Science* 27 (1960) 75–87, and: Geometry, Chronometry and Empiricism, Section 7, *Minnesota Studies in the Philosophy of Science* 3 (ed. by H. Feigl and G. Maxwell), University of Minnesota Press, Minneapolis, 1962.

whose verification or refutation is at issue in the conduct of particular scientific experiments, then Northrop's formal schema is a misleading oversimplification. Upon taking cognizance of the fact that the observational consequences O are deduced *not* from H alone but rather from the conjunction of H and the relevant body of *auxiliary* assumptions A, the refutability of H is seen to be no more conclusive than its verifiability. For now it appears that Northrop's formal schema must be replaced by the following:

$$(1) \quad [(H \cdot A) \to O] \cdot O \qquad \text{(verification)}$$

and

$$(2) \quad [(H \cdot A) \to O] \cdot \sim O \qquad \text{(refutation)}.$$

The recognition of the presence of the auxiliary assumptions A in both the verification and refutation of H now makes apparent that the *refutation* of H *itself* by *adverse* empirical evidence $\sim O$ can be no more decisive than its *verification* (confirmation) by *favorable* evidence O. What can be inferred deductively from the refutational premise (2) is *not* the falsity of H itself but only the much weaker conclusion that H and A cannot both be true. It is immaterial here that the *falsity* of the *conjunction* of H and A can be inferred *deductively* from the refutational premise (2) while the truth of that conjunction can be inferred only *inductively* from the verificational premise (1). For this does *not* detract from the fact that there is parity between the refutation of H *itself* and the verification of H itself in the following sense: (2) does *not* entail (deductively) the falsity of H itself, just as (1) does not entail the truth of H by itself. In short, isolated component hypotheses of far-flung theoretical systems are not separately refutable but only contextually disconfirmable. And Northrop's schema is an adequate representation of the actual logical situation only if "T_1" in his schema refers to the entire theoretical *system* of premises which enters into the deduction of O rather than to such mere *components* H as are at issue in specific scientific inquiries.

Under the influence of Duhem's emphasis on the confrontation of an entire theoretical system by the tribunal of evidence, writers such as W. v. O. Quine have gone further to make what I take to be the following claim: no matter what the specific content O' of the *prima facie* adverse empirical evidence $\sim O$, we can always justifiably affirm the truth of H as part of the theoretical *explanans* of O' by doing two things: (1) blame the falsity of O on the falsity of A rather than on the falsity of H, and

(2) so modify *A* that the conjunction of *H* and the *revised* version *A'* of *A* does entail (explain) the actual findings *O'*. Thus, in his *Two Dogmas of Empiricism*, Quine writes: "Any statement can be held true come what may, if we make drastic enough adjustments elsewhere in the system."[3] And one of Quine's arguments in that provocative essay against the tenability of the analytic-synthetic distinction is that a supposedly synthetic statement, no less than a supposedly analytic one can be claimed to be true "come what may" on Duhemian grounds.

The aim of my present paper is to establish two main conclusions:

(I) Quine's formulation of Duhem's thesis – hereafter called the "*D*-thesis" – is true *only* in various *trivial* senses of what Quine calls "drastic enough adjustments elsewhere in the system". And no one would wish to contest any of these thoroughly uninteresting versions of the *D*-thesis,

(II) in its *non*-trivial, exciting form, the *D*-thesis is untenable in the following fundamental respects:

A. *Logically*, it is a *non-sequitur*. For *independently* of the particular empirical context to which the hypothesis *H* pertains, there is no logical guarantee at all of the existence of the *required kind* of revised set *A'* of auxiliary assumptions such that

$$(H \cdot A') \to O'$$

for any one component hypothesis *H* and any *O'*. Instead of being guaranteed logically, the existence of the required set *A'* needs *separate* and *concrete* demonstration for each particular context. In the absence of the latter kind of *empirical* support for Quine's unrestricted Duhemian claim, that claim is an unempirical dogma or article of faith which the pragmatist Quine is no more entitled to espouse than an empiricist would be.

B. The *D*-thesis is not only a *non-sequitur* but is actually *false*, as shown by an important counter-example, namely the *separate* falsifiability of a particular component hypothesis *H*.

To forestall misunderstanding, let it be noted that my rejection of the very strong assertion made by Quine's *D*-thesis is *not* at all intended as a repudiation of the following far weaker contention, which I believe to be eminently sound: the logic of every disconfirmation, no less than of every confirmation of an isolated scientific hypothesis *H* is such as to

180

involve at some stage or other an entire network of interwoven hypotheses in which H is ingredient rather than in every stage merely the separate hypothesis H. Furthermore, it is to be understood that the issue before us is the *logical* one whether *in principle* every component H is unrestrictedly preservable by a suitable A', *not* the *psychological* one whether scientists possess sufficient ingenuity at every turn to propound the required set A', *if it exists*. Of course, *if* there are cases in which the requisite A' simply *does not even exist logically*, then surely no amount of ingenuity on the part of scientists will enable them to ferret out the non-existent required A' in such cases.

I. THE TRIVIAL VALIDITY OF THE D-THESIS

It can be made evident at once that unless Quine restricts in very specific ways what he understands by "drastic enough adjustments elsewhere in the (theoretical) system" the *D*-thesis is a thoroughly unenlightening truism. For if someone were to put forward the false empirical hypothesis H that "Ordinary buttermilk is highly toxic to humans", this hypothesis could be saved from refutation in the face of the observed wholesomeness of ordinary buttermilk by making the following "drastic enough" adjustment in our system: changing the rules of English usage so that the intension of the term "ordinary buttermilk" is that of the term "arsenic" in its customary usage. Hence a *necessary* condition for the non-triviality of Duhem's thesis is that *the theoretical language* be *semantically stable* in the relevant respects.

Furthermore, it is clear that if one *were* to countenance that O' itself qualifies as A', Duhem's affirmation of the existence of an A' such that

$$(H \cdot A') \rightarrow O'$$

would hold trivially, and H would not even be needed to deduce O'. Moreover, the *D*-thesis can hold trivially even in cases in which H is required in addition to A' to deduce the *explanandum* O': an A' of the trivial form

$$\sim H \vee O'$$

requires H for the deduction of O', but no one will find it enlightening to be told that the *D*-thesis can thus be sustained.

I am unable to give a formal *and* completely general *sufficient* condition for the *non*-triviality of A'. And, so far as I know, neither the origi-nator nor any of the advocates of the D-thesis have even shown any awareness of the need to circumscribe the class of *non*-trivial revised auxiliary hypotheses A' so as to render the D-thesis interesting. I shall therefore assume that the proponents of the D-thesis intend it to stand or fall on the kind of A' which we would all recognize as *non*-trivial in *any given case*, a kind of A' which I shall symbolize by "A'_{nt}". And I shall endeavour to show that such a *non*-trivial form of the D-thesis is indeed untenable after first commenting on the attempt to sustain the D-thesis by resorting to the use of a *non-standard logic*.

The species of drastic adjustment consisting in recourse to a *non*-standard logic is specifically mentioned by Quine. Citing a hypothesis such as "there are brick houses on Elm Street", he claims that even a statement so "germane to sense experience... can be held true in the face of recalcitrant experience by pleading hallucination or by amending certain statements of the kind called logical laws". (*Ibid.*, p. 43). I dis-regard for now the argument from hallucination. In the absence of *specifics* as to the ways in which alterations of logical laws will enable Quine to hold in the face of *recalcitrant* experience that a statement H like "there are brick houses on Elm Street" is *true*, I must conclude the following: the invocation of non-standard logics either makes the D-thesis *trivially* true or turns it into an interesting claim which is an unfounded dogma. For suppose that the non-standard logic used is a 3-valued one. Then even if it were otherwise feasible to assert within the framework of such a logic that the particular statement H is "true", the term "true" would no longer have the meaning associated with the 2-valued framework of logic within which the D-thesis was enunciated to begin with. It is not to be overlooked that a form of the D-thesis which allows itself to be sustained by alterations in the meaning of "true" is no less trivial *in the context of the expectations raised by the D-thesis* than one which rests its case on calling arsenic "buttermilk". And this triviality obtains *in this context*, notwithstanding the fact that the 2-valued and 3-valued usages of the word "true" share what H. Putnam has usefully termed a common "core meaning".[4] For suppose we had two particular substances I_1 and I_2 which are isomeric with each other. That is to say, these substances are composed of the same elements in the same propor-tions and with the same molecular weight but the arrangement of the atoms within the molecule is different. Suppose further that I_1 is not at all toxic while I_2 is highly toxic, as in the case of two isomers of trinitro-benzene.[5] Then if we were to call I_1 "duquine" and asserted that "duquine is highly toxic", this statement H could also be trivially saved from refutation in the face of the evidence of the wholesomeness of I_1 by the following device: only *partially* changing the meaning of "duquine" so that its intension is the second, highly toxic isomer I_2, thereby leaving the chemical "core meaning" of "duquine" intact. To avoid misunderstanding of my charge of triviality, let me point out precisely what I regard as trivial here. The preservation of H from refutation in the face of the evidence by a *partial* change in the meaning of "duquine" is trivial in the sense of being only a *trivial* fulfillment of *the expectations raised by the D-thesis*. But, in my view, the possibility *as such* of preserving H by *this particular kind of change in meaning* is not at all trivial. For this possibility as such reflects a fact about the world: the existence of isomeric substances of radically different degrees of toxicity (allergenicity)!

Even if one ignores the change in the meaning of "true" inherent in the resort to a 3-valued logic, there is no reason to think that the D-thesis can be successfully upheld in such an altered logical framework: the arguments which I shall present against the *non*-trivial form of the D-thesis within the framework of the standard logic apply just as much, so far as I can see, in the 3-valued and other non-standard logics of which I am aware. And if the reply is that there are *other* non-standard logics which are both viable for the purposes of science and in which my impending polemic against the non-trivial form of the D-thesis does *not* apply, then I retort: as it stands, Quine's assertion of the feasibility of a change in the laws of logic which would thus sustain the D-thesis is an unempirical dogma or at best a promissory note. And until the requisite collateral is supplied, it is not incumbent upon anyone to accept that promissory note.

II. THE UNTENABILITY OF THE NON-TRIVIAL D-THESIS

A. The Non-Trivial D-Thesis is a Non-Sequitur

The non-trivial D-thesis is that for every component hypothesis H of any domain of empirical knowledge and for any observational findings O',

$$(\exists A'_{nt}) [(H \cdot A'_{nt}) \to O'] .$$

But this claim does *not* follow from the fact that the falsity of *H* is *not* deductively inferable from premise (2) above, i.e., from

$$[(H \cdot A) \to O] \cdot \sim O .$$

For the latter premise utilizes *not* the full empirical information given by O' but only the part of that information which tells us that O' is logically incompatible with O. Hence the *failure* of $\sim O$ to permit the deduction of $\sim H$ does *not* justify the assertion of the *D*-thesis that there always *exists* a non-trivial A' such that the conjunction of *H* and that A' entails O'. In other words, the fact that the falsity of *H* is *not* deducible (by *modus tollens*) from premise (2) is quite insufficient to show that *H* can be preserved non-trivially as part of an *explanans* of *any* potential empirical findings O'. I conclude, therefore, from the analysis given so far that in its *non*-trivial form, Quine's *D*-thesis is *gratuitous* and that the existence of the required non-trivial A' would require *separate* demonstration for each particular case.

B. *Physical Geometry as a Counter-Example to the Non-Trivial D-Thesis*

Einstein has articulated Duhem's claim by reference to the special case of testing a hypothesis of physical geometry. In opposition to the Carnap-Reichenbach conception, Einstein maintains [6] that no hypothesis of physical geometry by itself is falsifiable even though all of the terms in the vocabulary of the geometrical theory, including the term "congruent" for line segments and angles, have been given a specific physical interpretation. And the substance of his argument is briefly the following: In order to follow the practice of ordinary physics and use rigid solid rods as the physical standard of congruence in the determination of the geometry, it is essential to make computational allowances for the thermal, elastic, electromagnetic, and other deformations exhibited by solid rods. The introduction of these corrections is an essential part of the logic of testing a physical geometry. For the presence of inhomogeneous thermal and other such influences issues in a dependence of the coincidence behaviour of transported solid rods on the latter's *chemical composition*. Now, Einstein argues that the geometry itself can never be accessible to experimental falsification *in isolation from* those other laws of physics which enter into the calculation of the corrections compensating for the distortions of the rod. And from this he then concludes that you can always preserve any geometry you like by suitable adjustments in the associated correctional physical laws. Specifically, he states his case in the form of a dialogue in which he attributes his own Duhemian view to Poincaré and offers that view in opposition to Hans Reichenbach's conception. But I submit that Poincaré's text will *not* bear Einstein's interpretation. For in speaking of the variations which solids exhibit under distorting influences, Poincaré says "we neglect these variations in laying the foundations of geometry, because, besides their being very slight, they are irregular and consequently seem to us accidental".[7] I am therefore taking the liberty of replacing the name "Poincaré" in Einstein's dialogue by the term "Duhem and Einstein". *With this modification,* the dialogue reads as follows:

"Duhem and Einstein: The empirically given bodies are not rigid, and consequently can not be used for the embodiment of geometric intervals. Therefore, the theorems of geometry are not verifiable.

Reichenbach: I admit that there are no bodies which can be *immediately* adduced for the "real definition" (i.e. physical definition) of the interval. Nevertheless, this real definition can be achieved by taking the thermal volume-dependence, elasticity, electro- and magneto-striction, etc., into consideration. That this is really and without contradiction possible, classical physics has surely demonstrated.

Duhem and Einstein: In gaining the real definition improved by yourself you have made use of physical laws, the formulation of which presupposes (in this case) Euclidean geometry. The verification, of which you have spoken, refers, therefore, not merely to geometry but to the entire system of physical laws which constitute its foundation. An examination of geometry by itself is consequently not thinkable. – Why should it consequently not be entirely up to me to choose geometry according to my own convenience (i.e., Euclidean) and to fit the remaining (in the usual sense "physical") laws to this choice in such manner that there can arise no contradiction of the whole with experience?"

By speaking here of the "real definition" (i.e., the coordinative definition) of "congruent intervals" by the corrected transported rod, Einstein is ignoring that the actual and potential physical meaning of congruence in physics *cannot* be given exhaustively by any *one* physical criterion or test condition. But here we can safely ignore this open cluster character of the concept of congruence. For our concern as well as Einstein's is

185

merely to single out *one* particular congruence class from among an infinitude of such alternative classes. And as long as our specification of that one chosen class is *unambiguous*, it is wholly immaterial that there are also *other* physical criteria (or test conditions) by which it could be specified.

Einstein is making two major points here: (1) In obtaining a physical geometry by giving a physical interpretation of the postulates of a formal geometric axiom system, the specification of the physical meaning of such theoretical terms as "congruent", "length", or "distance" is *not* at all simply a matter of giving an operational definition in the strict sense. Instead, what has been variously called a "rule of correspondence" (Margenau and Carnap), a "coordinative definition" (Reichenbach), an "epistemic correlation" (Northrop), or a "dictionary" (N. R. Campbell) is provided here *through the mediation of hypotheses and laws* which are collateral to the geometric theory whose physical meaning is being specified. Einstein's point that the physical meaning of congruence is given by the transported rod *as corrected theoretically* for idiosyncratic distortions is an illuminating one and has an abundance of analogues throughout physical theory, thus showing, incidentally, that strictly operational definitions are a rather simplified and limiting species of rules of correspondence. In particular, we see that the physical interpretation of the term "length", which is often adduced as the prototype of all "operational" definitions in Bridgman's sense, is *not* given operationally in any *distinctive* sense of that ritually invoked term. (2) Einstein's second claim, which is the cardinal one for our purposes, is that the role of collateral theory in the physical definition of congruence is such as to issue in the following *circularity*, from which there is no escape, he maintains, short of acknowledging the existence of an a priori element *in the sense of the Duhemian ambiguity:* the rigid body is not even defined without first *decreeing* the validity of Euclidean geometry (or of some other particular geometry). For *before* the *corrected* rod can be used to make an empirical determination of the *de facto* geometry, the required corrections must be computed via laws, such as those of elasticity, which involve Euclideanly calculated areas and volumes.[8] But clearly the warrant for thus introducing Euclidean geometry *at this stage* cannot be empirical.

In the same vein, H. Weyl endorses Duhem's position as follows: "Geometry, mechanics, and physics form an inseparable theoretical

186

whole [9] ...Philosophers have put forward the thesis that the validity or non-validity of Euclidean geometry cannot be proved by empirical observations. It must in fact be granted that in all such observations essentially physical assumptions, such as the statement that the path of a ray of light is a straight line and other similar statements, play a prominent part. This merely bears out the remark already made above that it is only the whole composed of geometry and physics that may be tested empirically." [10]

I now wish to set forth my doubts regarding the soundness of Einstein's geometrical form of the D-thesis. [11] And I shall do so in two parts the first of which deals with the special case in which effectively no deforming influences are present in a certain region whose geometry is to be ascertained.

If we are confronted with the problem of the falsifiability of the geometry ascribed to a region which is effectively free from deforming influences, then the *correctional* physical laws play no role as auxiliary assumptions, and the latter reduce to the claim that the region in question is, in fact, effectively *free* from deforming influences. And *if* such freedom can be affirmed *without* presupposing collateral theory, then the geometry alone rather than only a wider theory in which it is ingredient will be falsifiable. On the other hand, if collateral theory *were* presupposed here, then Duhem and Einstein might be able to adduce its modifiability to support their claim that the geometry *itself* is *not* separately falsifiable.

Specifically, they might argue then that the collateral theory could be modified such that the region then turns out *not* to be free from deforming influences with resulting inconclusive falsifiability of the geometry. The question is therefore whether freedom from deforming influences can be asserted and ascertained independently of (sophisticated) collateral theory. My answer to this question is Yes. For quite independently of the conceptual elaboration of such physical magnitudes as temperature, whose constancy would characterize a region free from deforming influences, the absence of perturbations is certifiable for the region as follows: two solid rods of very *different* chemical constitution which coincide at one place in the region will also coincide everywhere else in it (independently of their paths of transport). It would *not* do for the Duhemian to object here that the certification of two solids as quite *different chemically* is theory-laden to an extent permitting him to uphold his thesis of the inconclusive falsifiability of the geometry. For suppose

187

that observations were so ambiguous as to permit us to assume that two solids which appear strongly to be chemically *different* are, in fact, chemically identical in all relevant respects. If so rudimentary an observation were thus ambiguous, then no observation could ever possess the required univocity to be incompatible with an observational consequence of a *total theoretical* system. And if that were the case, Duhem could hardly avoid the following conclusion: "observational findings are always so unrestrictedly ambiguous as not to permit even the refutation of any given *total theoretical* system." But such a result would be tantamount to the absurdity that *any* total theoretical system can be espoused as true a priori. By the same token, incidentally, I cannot see what methodological safeguards would prevent Quine from having to countenance such an outcome within the framework of his *D*-thesis. In view of his avowed willingness to "plead hallucination" to deal with observations *not* conforming to the hypothesis that "there are brick houses on Elm Street", one wonders whether he would be prepared to say that *all* human observers who make disconfirming observations on Elm Street are hallucinating. And, if so, why not discount all observations incompatible with an *arbitrary total theoretical system* as hallucinatory? Thus, it would seem that if Duhem is to maintain, as he does, that a *total theoretical system is* refutable by confrontation with observational results, then he must allow that the coincidence of diverse kinds of rods at different places in the region (independently of their paths of transport) is certifiable observationally. Accordingly, the absence of deforming influences is ascertainable *independently* of any assumptions as to the geometry and of other (sophisticated) collateral theory.

Let us now employ our earlier notation and denote the geometry by "H" and the assertion concerning the freedom from perturbations by "A". Then, once we have laid down the congruence definition and the remaining semantical rules, the physical geometry H becomes *separately* falsifiable as an *explanans* of the posited empirical findings O'. It is true, of course, that A is only more or less highly confirmed by the ubiquitous coincidence of chemically different kinds of solid rods. But the inductive risk thus inherent in affirming A does *not* arise from the alleged inseparability of H and A, and that risk can be made exceedingly small without any involvement of H. Accordingly, the actual logical situation is characterized *not* by the Duhemian schema but instead by the schema

188

$$[\{(H \cdot A) \to O\} \cdot \sim O \cdot A] \to \sim H .$$

We now turn to the critique of Einstein's Duhemian argument as applied to the empirical determination of the geometry of a region which is subject to deforming influences.

There can be no question that when deforming influences *are* present, the laws used to make the corrections for deformations involve areas and volumes in a fundamental way (e.g., in the definitions of the elastic stresses and strains) and that this involvement presupposes a geometry, as is evident from the area and volume formulae of differential geometry, which contains the square root of the determinant of the components g_{ik} of the metric tensor.[12] Thus, the empirical determination of the geometry involves the joint assumption of a geometry and of certain collateral hypotheses. But we see already that this assumption *cannot* be adequately represented by the conjunction $H \cdot A$ of the Duhemian schema, where H represents the geometry.

Now suppose that we begin with a set of Euclideanly formulated physical laws P_0 in correcting for the distortions induced by perturbations and then use the thus Euclideanly corrected congruence standard for *empirically* exploring the geometry of space by determining the metric tensor. *The initial stipulational affirmation of the Euclidean geometry G_0 in the physical laws P_0 used to compute the corrections in no way assures that the geometry obtained by the corrected rods will be Euclidean!* If it is *non*-Euclidean, then the question is: What will be required by Einstein's fitting of the physical laws to preserve Euclideanism and avoid a contradiction of the theoretical system with experience? Will the adjustments in P_0 necessitated by the retention of Euclidean geometry entail merely a change in the dependence of the length assigned to the transported rod on such *nonpositional* parameters as temperature, pressure, and magnetic field? Or could the putative empirical findings compel that the length of the transported rod be likewise made a nonconstant function of its *position* and *orientation* as *independent* variables in order to square the coincidence findings with the requirement of Euclideanism? The possibility of obtaining *non*-Euclidean results by measurements carried out in a spatial region uniformly characterized by standard conditions of temperature, pressure, electric and magnetic field strength, etc., shows it to be *extremely doubtful*, as we shall now show, that the preservation of

189

Euclideanism could *always* be accomplished short of introducing *the dependence of the rod's length on the independent variables of position or orientation.*

But the introduction of the latter dependence is none other than so radical a change in the meaning of the word "congruent" that this term now denotes a class of intervals *different* from the original congruence class denoted by it. And such tampering with the semantical anchorage of the word "congruent" violates the requirement of semantical stability, which is a necessary condition for the *non*-triviality of the *D*-thesis, as we saw above.

Suppose that, relatively to the customary congruence standard, the geometry prevailing in a given region when *free* from perturbational influences is that of a strongly *non*-Euclidean space of spatially and temporally constant curvature. Then what would be the character of the alterations in the *customary* correctional laws which Einstein's thesis would require to assure the *Euclideanism* of that region relatively to the customary congruence standard under *perturbational* conditions? The required alterations would be *independently falsifiable*, as will now be demonstrated, because they would involve affirming that such coefficients as those of linear thermal expansion *depend on the independent variables of spatial position*. That such a space dependence of the correctional coefficients might well be necessitated by the exigencies of Einstein's Duhemian thesis can be seen as follows by reference to the law of linear thermal expansion. In the usual version of physical theory, the first approximation of that law [13] is given by

$$L = L_0(1 + \alpha \cdot \Delta T).$$

If Einstein is to guarantee the Euclideanism of the region under discussion by means of logical devices that are consonant with his thesis, and if our region is subject only to *thermal* perturbations for some time, then we are confronted with the following situation: unlike the customary law of linear thermal expansion, the revised form of that law needed by Einstein will have to bear the *twin* burden of effecting *both* of the following two kinds of superposed corrections: (1) the *changes* in the lengths ascribed to the transported rod in different positions or orientations which would be required even if our region *were* everywhere at the standard tempera-

190

ture, merely for the sake of rendering Euclidean its otherwise *non-*Euclidean geometry, and (2) corrections compensating for the effects of the *de facto* deviations from the standard temperature, *these* corrections being the *sole onus* of the *usual* version of the law of linear thermal expansion. What will be the consequences of requiring the *revised* version of the law of thermal elongation to implement the *first* of these two kinds of corrections in a context in which the deviation ΔT from the standard temperature is the *same* at *different* points of the region, that temperature deviation having been measured in the manner chosen by the Duhemian? Specifically, what will be the character of the coefficients α of the *revised* law of thermal elongation under the posited circumstances, if Einstein's thesis is to be implemented by effecting the *first* set of corrections? Since the new version of the law of thermal expansion will then have to guarantee that the lengths L assigned to the rod at the various points of *equal* temperature T *differ* appropriately, it would seem clear that logically possible empirical findings could compel Einstein to make the coefficients α of solids *depend* on the *space coordinates*.

But such a spatial dependence is *independently falsifiable:* comparison of the thermal elongations of an aluminum rod, for example, with an invar rod of essentially zero α by, say, the Fizeau method *might* well show that the α of the aluminum rod is a characteristic of aluminum which is *not* dependent on the space coordinates. And even if it *were* the case that the α's are found to be space dependent, how could Duhem and Einstein assure that this space dependence would have the particular functional form required for the success of their thesis?

We see that the required resort to the introduction of a spatial dependence of the thermal coefficients might well *not* be open to Einstein. Hence, in order to retain Euclideanism, it would then be necessary to *remetrize* the space in the sense of abandoning the customary definition of congruence, entirely apart from any consideration of idiosyncratic distortions and even after correcting for these in some way or other. But this kind of remetrization, though entirely admissible in *other* contexts, does *not* provide the requisite support for Einstein's Duhemian thesis! For Einstein offered it as a criticism of Reichenbach's conception. And hence it is the *avowed onus* of that thesis to show that the geometry by *itself* cannot be held to be empirical, i.e., separately falsifiable, even when, with Reichenbach, we have sought to assure its empirical character by choosing and then adhering to the usual (standard) definition of spatial congruence, which *excludes* resorting to such remetrization.

Thus, there may well obtain observational findings O', expressed in terms of a particular definition of congruence (e.g., the *customary* one), which are such that there does *not* exist any nontrivial set A' of auxiliary assumptions capable of preserving the Euclidean H in the face of O'. And this result alone suffices to invalidate the Einsteinian version of Duhem's thesis to the effect that any geometry, such as Euclid's, can be preserved in the face of any experimental findings which are expressed in terms of the customary definition of congruence.

It might appear that my geometric counterexample to the Duhemian thesis of unavoidably contextual falsifiability of an *explanans* is vulnerable to the following criticism: "To be sure, Einstein's geometric articulation of that thesis does not leave room for saving it by resorting to a remetrization in the sense of making the length of the rod *vary* with position or orientation even *after* it has been corrected for idiosyncratic distortions. But why saddle the Duhemian thesis as such with a restriction peculiar to Einstein's particular version of it? And thus why not allow Duhem to save his thesis by countenancing those *alterations in the congruence definition* which are *remetrizations*?"

My reply is that to deny the Duhemian the invocation of such an alteration of the congruence definition *in this context* is *not* a matter of gratuitously requiring him to justify his thesis within the confines of Einstein's particular version of that thesis; instead, the imposition of this restriction is entirely legitimate here, and the Duhemian could hardly wish to reject it as unwarranted. For it is of the essence of Duhem's contention that H (in this case, Euclidean geometry) can always be preserved *not* by tampering with the principal semantical rules (interpretive sentences) linking H to the observational base (i.e., specifying a particular congruence class of intervals etc.), but rather by availing oneself of the alleged *inductive latitude* afforded by the ambiguity of the experimental evidence to do the following: (a) leave the factual commitments of H *essentially unaltered* by retaining both the statement of H and the *principal* semantical rules linking its terms to the observational base, and (b) replace the set A by A' such that A and A' are logically incompatible under the hypothesis H. The qualifying words "principal" and "essential" are needed here in order to obviate the possible objection that it may not be logically possible to supplant the auxiliary assumptions A by A' *without also* changing the factual content of H in *some* respect. Suppose for example, that one were to abandon the optical hypothesis A that light will require equal times to traverse *congruent* closed paths in an inertial system in favor of some rival hypothesis. Then the semantical linkage of the term "congruent space intervals" to the observational base is changed to the extent that this term no longer denotes intervals traversed by light in equal round-trip times. But such a change in the semantics of the word "congruent" is innocuous in this context, since it leaves wholly intact the membership of the class of spatial intervals that is referred to as a "congruence class". In this sense, then, the modification of the optical hypothesis leaves intact both the *"principal"* semantical rules governing the term "congruent" *and* the *"essential"* factual content of the geometric hypothesis H, which is predicated on a particular congruence class of intervals. That "essential" factual content is the following: relatively to the congruence specified by unperturbed transported rods – *among other things* – the geometry is Euclidean.

Now, the essential factual content of a geometrical hypothesis can be changed either by preserving the original statement of the hypothesis while changing one or more of the principal semantical rules or by keeping all of the semantical rules intact and suitably changing the statement of the hypothesis. We can see, therefore, that the retention of a Euclidean H by the device of changing through remetrization the semantical rule governing the meaning of "congruent" (for line segments) effects a retention *not* of the essential *factual commitments* of the original Euclidean H but only of its *linguistic trappings*. That the thus "preserved" Euclidean H actually repudiates the essential factual commitments of the *original* one is clear from the following: the *original* Euclidean H had asserted that the coincidence behavior common to all kinds of solid rods is Euclidean, *if* such transported rods are taken as the physical realization of congruent intervals; but the Euclidean H which survived the confrontation with the posited empirical findings only by dint of a *remetrization* is predicated on a denial of the very assertion that was made by the original Euclidean H, which it was to "preserve". It is as if a physician were to endeavor to "preserve" an a priori diagnosis that a patient has acute appendicitis in the face of a negative finding (yielded by an exploratory operation) as follows: he would redefine "acute appendicitis" to denote the healthy state of the appendix!

Hence, the confines within which the Duhemian must make good his claim of the preservability of a Euclidean H do *not* admit of the kind of change in the congruence definition which alone would render his claim tenable under the assumed empirical conditions. Accordingly, the geometrical critique of Duhem's thesis given in this paper does *not* depend for its validity on restrictions peculiar to Einstein's version of it.

Even apart from the fact that Duhem's thesis precludes resorting to an alternative metrization to save it from refutation in our geometrical context, the very feasibility of alternative metrizations is vouchsafed *not* by any general Duhemian considerations pertaining to the logic of falsifiability but by a property peculiar to the subject matter of geometry (and chronometry): the latitude for *convention* in the ascription of the spatial (or temporal) *equality* relation to intervals in the continuous manifolds of physical space (or time).

It would seem that the least we can conclude from the analysis of Einstein's geometrical *D*-thesis given in this paper is the following: since empirical findings can greatly narrow down the range of uncertainty as to the prevailing geometry, there is no assurance of the *latitude* for the choice of a geometry which Einstein takes for granted in the manner of the *D*-thesis.

University of Pittsburgh, Pittsburgh, Pennsylvania

NOTES

1. F. S. C. Northrop, *The Logic of the Sciences and the Humanities*, New York, 1947, p. 146.
2. Cf. Pierre Duhem, *The Aim and Structure of Physical Theory*. Princeton, 1954, Part 2, Ch. 6, esp. pp. 183–190.
3. W. V. O. Quine, *From a Logical Point of View*, 2. ed., Cambridge, 1961, p. 43. Cf. also p. 41, n. 17.
4. H. Putnam, Three-Valued Logic, *Philosophical Studies*, 8 (1957) 74.
5. Cf. H. L. Alexander, *Reactions With Drug Therapy*, Philadelphia, 1955, p. 14. Alexander writes: "It is true that drugs with closely related chemical structures do not always behave clinically in a similar manner, for antigenicity of simple chemical compounds may be changed by minor alterations of molecular structures. ... 1, 2, 4-trinitrobenzene... is a highly antigenic compound. ...1, 3, 5... trinitrobenzene is allergenically inert." (I am indebted to Dr. A. I. Braude for this reference).

6. Cf. A. Einstein, Reply to Criticisms, *Albert Einstein: Philosopher-Scientist* (ed. by P. A. Schilpp). Tudor, New York, 1949, pp. 676–678.
7. H. Poincaré, *The Foundations of Science*, Lancaster, 1946, p. 76.
8. Cf. I. S. Sokolnikoff, *Mathematical Theory of Elasticity*. McGraw-Hill, New York, 1946. S. Timoshenko and J. N. Goodier, *Theory of Elasticity*. McGraw-Hill New York, 1951.
9. H. Weyl, *Space-Time-Matter*. Dover Publications, New York, 1950, p. 67.
10. *Ibid.*, p. 93.
11. I draw here on my more detailed treatment of this and related issues in A. Grünbaum, Geometry, Chronometry and Empiricism, *Minnesota Studies in the Philosophy of Science*, Vol. 3, Minneapolis, 1962, pp. 510–521.
12. L. P. Eisenhart, *Riemannian Geometry*. Princeton University Press, Princeton, 1949, p. 177.
13. This law is only the first approximation, because the rate of thermal expansion varies with the temperature. The general equation giving the magnitude m_t (length or volume) at a temperature t, where m_0 is the magnitude at 0° C, is

$$m_t = m_0(1 + \alpha t + \beta t^2 + \gamma t^3 + \ldots),$$

where α, β, γ, etc, are empirically determined coefficients (Cf. *Handbook of Chemistry and Physics*, Cleveland, 1941, p. 2194). The argument which is about to be given by reference to the approximate form of the law can be readily generalized to forms of the law involving more than one coefficient of expansion.

COMMENTS

ROBERT S. COHEN

What are the positive contributions of *both* Grünbaum and Duhem? This is not a contradiction since Grünbaum has on several significant occasions in the last few years stressed that Duhem has two things to say: one of them is both historically correct and a great insight, namely what Grünbaum referred to a couple of years ago as the weaker thesis of Duhem; the other put forth in somewhat unclear fashion by Duhem himself is the so-called stronger thesis of the impossibility of refuting any particular H, and it is this thesis which is incorrect.

We learned a number of things this evening. First, and very important, that pragmatism is still a devil to exorcize from our midst. There is more than just philosophy of science involved, there's a general epistemological issue here, and I think we should encourage our friend Grünbaum to give us a general philosophical article on the special, typical, and necessary Duhemian weakness in pragmatism. Related to this, I take it, or another way of stating this general comment, we learned tonight, and from the previous articles on the subject from Grünbaum, of the crucial importance of semantic confusions in the discussion of empirical verification of hypotheses. Time and time again there is hidden in the technological vocabulary of stating alternate ways of verifying hypotheses a redefining of the terms. The crucial thing for an empiricist or materialist or an objectivist in discussing this issue with conventionalists or pragmatists is to keep one's eyes very closely on the meaning of terms so that one may be sure that what is being verified is what one started out discussing. This appeared several times in this paper this evening. I believe the discussion of the example from geometry is quite correct, but the same point (in the geometric case it is about the need for remetrizing), namely the occurrence of an unwarranted semantical shift in the argument, was alleged by Grünbaum in the somewhat simpler example of houses being either brick or not brick, taken from Quine.

Now I want to offer two questions which may deserve some kind of clarification from Grünbaum. Everyone agrees, I'm sure, that the weak hypothesis of Duhem is correct, namely that hypotheses involve many

196

auxiliary statements with them. The frequent example of looking at stars and offering astronomical theories which willy-nilly include theories of optics, of how the telescopes work, and general electro-magnetic theory, etc., shows that you're not testing, refuting, or giving probable evidence for any one hypothesis, say about planetary motions, but a hypothesis plus auxiliaries. There is not much to argue about that. But let us take another kind of hypothesis, with apparently a different career in the history of science. It does seem at times, that such an hypothesis as that of the Neutrino was offered as a shift in a theory, a shift in the auxiliary A. By shifting A one was able to preserve a certain hypothesis H about conservation of momentum and energy. Now to many physicists this appeared at the time as a confirming instance of the truth of the Poincaré-Duhem hypothesis. It seemed that one could preserve what appeared to be refuted by the data by shifting the auxiliary assumptions. The auxiliaries were changed by including a new assumption of the existence of an otherwise unverifiable particle and thereby casting doubt on the status of verifiability in scientific theories. It all turned out happily in the end, but at the time physicists were unable to give independent, empirical confirmation of the existence of such a particle. It then seemed outlandishly impossible to verify its existence. This new auxiliary was a Duhem-Poincaré way of saving a hypothesis, and a proper way. If one wanted to pay the price of postulating an almost massless and chargeless particle, then one could save the conservation law. It appears that one could save what one wants by paying the price of assuming the existence of something else. Now this sounded to physicists like the confirmation of Duhem's strong hypothesis. It sounds to me more plausible than Professor Quine's example of the street totally made of wooden houses. He wants to maintain via a general hallucinatory hypothesis, the thesis that these are brick houses despite all the evidence from observation by all the rest of us, including himself, of wooden houses. Now this crazy example seems to lend support to Grünbaum's argument, because all of us instinctively reject it, yet historically and logically, I don't see that it is different from the Neutrino hypothesis.

An important merit of this whole discussion tonight, should be clear. Grünbaum is using, or intends us to understand, such words as "logical", "logical basis for holding a certain hypothesis", in such a way that we must take quite seriously an inductive rather than a deductive validation.

Quine and others appear to be saying that we are not *deductively* forbidden from holding an arbitrary hypothesis. Grünbaum counters that this doesn't mean one may conclude anything whatsoever about scientific theories or their justifications, because all arguments, in empirical science, are inductively formulated. Hence one must offer an inductive formulation to support the Duhem strong hypothesis and not just the deductive example. Hence we are dealing with an existential quantifier in the strong Duhem hypothesis, and this is the main burden of the first logical proof Grünbaum offered, to the effect that there is a non-sequitur. Specifically, one cannot assume that there exist other alternative auxiliary statements to accompany H. Quine and Duhem wish to allow all kinds of possible auxiliaries, so long as we pay a sufficient conceptual price. But Grünbaum says we simply cannot, that in some cases there were no such hypotheses possible in the empirical situation. In other cases, you may have to require that the entire human race and all possible observers are hallucinating. Are we prepared to pay that price? Certainly not in an inductive spirit. And I wonder whether Quine and Duhem are prepared to pay that price. One must discuss existential quantification in discussing the Duhem hypothesis. It is unfortunate that in the sixty years, I think, since Duhem propounded this view only now this point has been made so clearly, but we're fortunate that it has been made here tonight.

Boston University, Department of Physics, Boston, Massachusetts

NOAM CHOMSKY

PERCEPTION AND LANGUAGE

SUMMARY OF ORAL PRESENTATION

Presented May 17, 1962

---

1. The classical problem of perception is to determine how an organism uses its knowledge, beliefs, and expectations in interpreting a sensory input. In the case of perception of language, we can formulate for separate study a particular instance of this general question, namely: how does a person bring his knowledge of his language, his intrinsic mature linguistic competence, to bear in assigning to a speech signal a structural description?

We can, in other words, pose the problem of specifying the characteristics of a "perceptual device" PD that takes as its input a signal S and gives as its output a "percept" P:

(1) $\qquad$ S → | PD | → P

Approaching general problems of perception from linguistics, we may ask: what conditions does language structure place on the character and functioning of a perceptual device. This investigation breaks down into three sub-studies, namely, the study of the input signal, of the internal structure of the perceptual device, and of the percept that we can regard, quite naturally, as the result of its functioning.

2. For the purposes of this discussion, I will assume that the signal can be described with any desired accuracy – in particular, that we can assume it to be given in phonetic transcription in terms of a universal phonetic theory that contains a universal alphabet. Thus we assume fixed a denumerable set of possible language signals, common to all languages, from which the utterances of each language are chosen.

3. We can assume that the device PD contains a recursive specification of a set of pairs (S, P), where S is a signal and P is a percept (or, to

use a more technical term, a "structural description") associated with it in the language in question, as well as a strategy for finding P given S. About the strategy, I have nothing to say, and I believe that nothing significant is known. About the recursive specification of pairs (S, P) (what I will henceforth call the "grammar" of the language of the perceiver), quite a bit is known, and it seems to me that what is known sets non-trivial conditions on the theory of perception.

4. Consider now the structural description that is the output of the device PD of (1). It is not, of course, directly observable. Rather we must attempt to develop a theory of structural descriptions of sentences on the basis of indirect evidence concerning what information is available to one who has identified and correctly understood an utterance of his language. From the mass of evidence of this sort, we must attempt to abstract underlying systems that play a role in the understanding of utterances, and that can be studied independently in some non-trivial way. It is obvious on a moment's reflection that many factors of diverse sorts interact to give what observable evidence we have concerning linguistic behavior, and that failure to distinguish these (as, e.g., when language is regarded as a mass of undifferentiated "dispositions to respond") condemns the entire enterprise to chaos and sterility. Of the several coherent subsystems concerning which some information is available, I will sketchily discuss here only two: the system of "rational (phonemic) spelling" and the system of grammatical relations. Thus let us consider what is involved in "identifying" a signal as a certain string of phonemes and in assigning to it a certain network of grammatical relations.

5. A "minimal pair" is a pair of utterances that differ in only one position, e.g., "pin"-"bin", "bin"-"ban", etc., but not "pin"-"ban". It is not difficult to convey this notion to an unsophisticated speaker. One obvious condition on phonemic representation must be that, in general, pairs that are perceptually minimal should differ correspondingly in phonemic representation. Thus, for example, the minimal pairs

(2) (i)   said – set (sed – set)
　　(ii) set – sat (set – sæt)
　　(iii) filler – feeler (filɼ – fīlɼ)
　　(iv) racer – razor (rēsɼ – rēzɼ)

might be represented, respectively, as in the parentheses in a perceptually adequate orthography.

6. Consider now the matter of representation of grammatical relations. In simple cases, the grammatical relations in utterances can be represented by labelled bracketing, or, equivalently, in a diagram such as (3).

(3) (a)   (b)

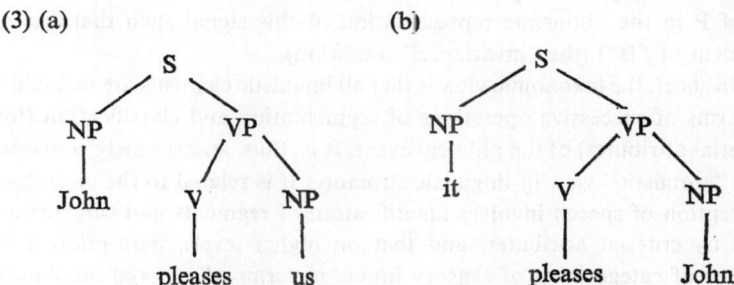

(3a) can be interpreted as indicating that "John" is a Noun phrase (NP) and "pleases us" a Verb phrase (VP) in "John pleases us," which is a sentence (S), while "pleases" is a Verb (V) and "us" a NP. Similarly, (3b). Grammatical relations can be defined in terms of subconfigurations of such "tree diagrams". Thus the Subject-Verb relation can be defined as the triple (S, NP, V), and we can say that this relation holds of the string tokens (x, y) in the string z with the structural description D if D contains the node S dominating NP to the left of VP, and if x is the terminal string dominated by this occurrence of NP in D, while y is the terminal string dominated by the main V of this occurrence of VP in D. There is no difficulty in making this notion precise, in one or another of the various natural ways.

7. The view concerning linguistic structure that has been dominant throughout the period when such questions have been seriously discussed is not unrelated to a certain rather elementary theory of perception. This view (call it the "taxonomic view") holds that grammatical relations are completely described in terms of labelled bracketing such as (3) (i.e., immediate constituent analysis, in one or another form) and that phonemic representation meets the following conditions: to each occurrence P of a phoneme in the phonemic representation of an utterance s, there corresponds a continuous segment $f(P)$ of the signal s – i.e., a sequence of one or more phones in its phonetic representation – and if P precedes Q, then

f (P) precedes f (Q) in s (the "linearity" condition); corresponding to each phoneme P, there is a set g (P) of "criterial attributes" such that if P′ is an occurrence of P in a phonemic representation, then f (P′) (in the sense just defined) is distinguished by the possession of g(P), and if a segment s of a signal is distinguished by possession of g(P), then there is an occurrence P′ of P in the phonemic representation of this signal such that s is a segment of f (P′) (the "invariance" condition).

In short, the taxonomic view is that all linguistic elements are definable in terms of successive operations of segmentation and classification (by criterial attributes) of the physical event. It is, thus, an extremely concrete and "atomistic" view of linguistic structure. It is related to the view that perception of speech involves identification of segments and categorization by criterial attributes, and that on higher levels, perception is a matter of categorizing of sensory inputs in terms of criterial attributes which, in some arrangement, define "concepts".

8. In the case of language, it is easily seen that the taxonomic conception is completely inadequate. Thus consider the following simple examples. On the level of phonemic representation, consider the expressions

(4) (a) said   [sed]   /sed/
          set   [set]   /set/
          sent   [sẽt]   /sent/
          send   [sẽnd]   /send/
     (b) writer [rayDṛ]   /raytṛ/
          rider   [ra·yDṛ]   /raydṛ/

where [...] encloses phonetic, and / ... / phonemic representations. Notice that in these examples both linearity and invariance are violated. Clearly it is not the case that (set-sent) constitute a minimal pair (with phonemic nasalization), while (sent-send) do not. Furthermore, it is clear that in case (4b), the phonemic difference is not vowel length (i.e., physically, writerrider is like (2iii), but intuitively is like (2iv)). But these decisions would be required by the taxonomic view. From numerous similar examples, it follows that perceptually appropriate phonemic representation is not based on segmentation and classification by criterial attributes. In fact, in each of these (and many other) cases, it can be shown that the phonemic representation is related to the phonetic by a sequence of ordered rules that do not, in general, preserve any simple relation

202

(such as linearity and invariance) between the perceptually appropriate phonemic representation and the physical event.

Turning briefly to the representation of grammatical relations, consider the pair of sentences with the labelled bracketing:

(5)

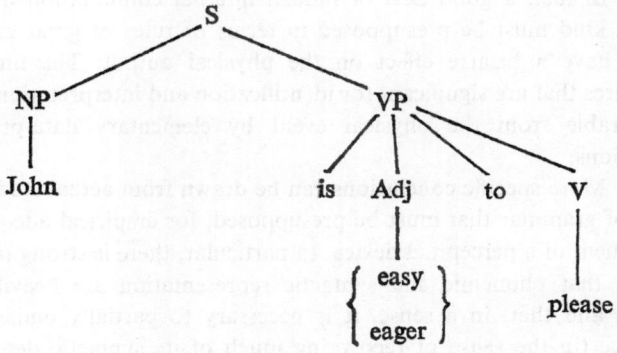

or something of this sort. Clearly the sentences

(6) John is easy to please

(7) John is eager to please

are not adequately represented by (5). No simple bracketing is capable of representing the grammatical relations in these sentences, namely, the fact that in (6) the relation between "John" and "please" is Object-Verb, while in (7) it is Subject-Verb. But clearly this is the formal basis for the way these sentences are understood. Again, we see that the taxonomic view is incorrect, and that interpreting a sentence is also not reducible to operations of segmentation and categorization. A detailed investigation of such examples as these, shows (just as in the case of phonemic representation, but much more drastically) that the underlying structures that are "perceptually appropriate" (i.e., that represent the information available to the person who has correctly understood the sentence) are remotely related to the physical event by formal operations, which, though highly systematic and general, are quite abstract.

9. In short, in order to account for both the identification (phonemic representation) and interpretation of a sentence (in particular, determination of the grammatical relations among its parts), it is necessary to dis-

card the taxonomic view, along with the elementary theory of perception that underlies it, and to consider a perceptual device incorporating a much more abstract and highly structured grammar which generates structural descriptions ("potential percepts") incorporating representations that have no direct relation, in general, to physical segments of the signal. In fact, a good deal of hidden internal computation of a very special kind must be presupposed in terms of rules of great generality which have a bizarre effect on the physical output. The underlying structures that are significant for identification and interpretation are not recoverable from the physical event by elementary data-processing operations.

10. More specific conclusions can be drawn from actual study of the form of grammar that must be presupposed, for empirical adequacy, as an element of a perceptual device. In particular, there is strong reason to believe that phonemic and syntactic representation are heavily inter-related and that, in a sense, it is necessary to partially understand a sentence (in the sense of recovering much of its syntactic description) before a full phonetic representation can be assigned to it. Investigation of transformational generative grammars, and the kinds of structural de-scriptions that they produce, suggests a two-stage processing of signals, the first of which determines the superficial structure (e.g., such rep-resentations as (5) for (6) and (7)), which is utilized in determining the phonetic representation (i.e., in fully identifying the utterance); while the second assigns deeper structural representations to the utterance (i.e., it would assign (3a) as a partial underlying representation for (7), and (3b) for (6)). The first component requires only a relatively small memory (for several reasons, in particular, because the node-to-terminal-node ratio, which gives a first approximation to the amount of "processing" required per input symbol, is typically small in superficial structures as (5), though it is typically high in such deeper representations as (3a) and (3b)), while the second would have to incorporate the full set of generative devices of the transformational grammar, and would require much larger memory.

11. For more details concerning these matters, see the three chapters by G. A. Miller and N. Chomsky in the forthcoming *Handbook of Mathematical Psychology* (to be published by Wiley – Luce, Galanter, Bush, editors), and the references cited there. The point I would like to

stress is that formal analysis of a complex product of human intellectual activity can impose non-trivial conditions that a perceptual theory must meet. Investigation of languages reveals underlying structures that are incompatible with elementary taxonomic views based on segmentation and classification by criterial attributes, just as it suggests that a great deal of internal organization, of a highly specific and intricate kind, may well be needed to account for the acquisition of such systems.

Massachusetts Institute of Technology, Cambridge, Massachusetts

205

SIDNEY MORGENBESSER

PERCEPTION: CAUSE AND ACHIEVEMENT

SUMMARY OF ORAL PRESENTATION

Presented May 17, 1962

Recently Ryle, Hamlyn, White and other analytic philosophers have directly or indirectly challenged the rationality of the psychologist's aim to construct explanatory theories of perception. Hamlyn has reviewed various psychological theories and programs in considerable detail, White and Ryle have not, at least not in print. But the criticisms of the latter of well known philosophical analyses of 'perception' apply and have been applied to psychological theories as well. These critiques raise diverse philosophical and metaphilosophical issues; I will attend only to those which bear upon the assessment of the psychologist's aim and try to defend it against its critics.

One version of the program maintained is that psychology aims to construct one theory of perception and then to reduce it to a physiological one. This is, of course, a grandiose quest whose chances for success are slight. But this, from our perspective, is of minor moment; we are considering not the ingenuity of the psychologist but the coherence of his program. Restricting ourselves to the matter at hand and to the program Hamlyn specified above, we need therefore consider only two criticisms, both emphasized by Hamlyn; one that there is no good reason to think that one theory of perception will suffice to account for all perceptual phenomena, the other that it is a mistake in principle to believe that a theory of perception is reducible to a physiological one.

Nothing much need be said about the first and nothing much can be said, at least here, about the second criticism. The first criticism raises no burning philosophical issue; paraphrasing Wittgenstein we may say that the number one plays no special role in science; the second criticism is not developed by Hamlyn. He insists that physiological conditions are only necessary but never sufficient for perception, but gives no independent argument in defense of this thesis. I presume he would claim that he need not be asked for any since he is ready to defend the stronger one that no causal explanation or causal theory of perception is possible. But though he is right, his silence on this matter is disappointing; there are special problems about reduction that should be faced directly.

Ryle, to whom Hamlyn appeals on this issue, is not explicit. He argues effectively against the thesis that statements of the form 'A proves X' are synonymous with statements of the form 'A's body is in state K' but has not, I think, directly considered the claim which would be sufficient for a reductionist program – that statements of the former kind are extensionally equivalent to suitable statements of the latter. His animadversion that "we have (mistakenly) yielded to the temptation to push the concepts of seeing, hearing and the rest through the sorts of hoops that are the proper ones for the concepts which belong to the sciences of optics, acoustics, physiology, etc."[1] But he adds that "there are all sorts of important connexions between the things which have been and will be discovered in the sciences of optics, acoustics, neurophysiology and the rest".[2] Whether the connections are strong enough to support a reduction is not further considered by Ryle.

The version of the psychological program considered above was not articulate on at least one important issue. It did not specify whether the laws or theories of perception would be statistical or deterministic ones. With this distinction a variety of formulations of the psychological program can be specified; but we shall consider only those which insist upon the construction of causal laws and theories. Statistical theories have not been considered by analytic philosophers; the oversight is not minor but I shall not dwell on it.

The strongest argument against deterministic theories is the one developed by White and others – that there are no sufficient conditions for perception. But there are a few observations by Ryle and Hamlyn which perhaps were intended to support the claim that it is a mistake to seek causal laws or causal explanations of perception; at any rate it is out of place to remind ourselves that a causal explanation of perceptual event E need only satisfy the familiar requirement that a statement describing that event is deduced from an appropriate law containing *explanans* which also contain statements describing initial conditions and events that occurred prior to E. It need not satisfy the additional requirement that the event which is explained is not a momentary one and does not extend in time. Hence nothing essentially follows for the issue at home from Ryle's important observation that "the verbs 'see' and 'hear' do not stand for anything that goes on, that it has a beginning, middle and end".[3] And since there is no restriction as to the type of laws or the type of events or conditions that can be specified in the *explanans*, we will not consider here Hamlyn's observation that we cannot explain why someone sees what he does see simply by referring to the stimulating conditions or sensory events that are occurring when the person is seeing, and that we must also refer to his "capacities and abilities".[4] To argue for a causal explanation of perception is not to argue that perception is an effect of stimulation in the way, let us say, the gushing of blood is an effect of a cut.

Nor does anyone who claims that a causal explanation of perception is possible argue that the term "perception" is definable behavioristically or deny that "perception" is an achievement term in the sense that a man who sees lives up to standards at least in the trivial sense, that he has successfully applied various concepts to that which is visually present. But I think it is a mistake to argue with Peters and others that it follows, therefore, that no causal explanation of perception and action in general is possible on the grounds that "there cannot therefore be a sufficient explanation of actions in causal terms because, as Popper has put it, there is a logical gulf between nature and convention. Statements implying norms and standards cannot be deduced from statements about mere movements which have no such normative implications".[5] But if Peters is right then dentistry is based on a conceptual confusion (since this argument would hold against pain). His mistake is patent; he is confusing the issue of the definability of "perception" with the one of giving a causal account of perception.

White's argument is similar but deserves fuller treatment. He reasons that "A causal analysis of perception, as opposed to 'sensing', seems to me to have the difficulty that whereas failures can have sufficient conditions, successes can have only necessary conditions. If I am prevented from seeing (that is, it is made impossible for me to see), then it follows that I do not see; but if I am enabled to see (that is, it is made possible for me to see), then it does not follow that I do see."[6]

As it stands the argument is subject to the obvious rejoinder that if I make it possible for a dog to move (by removing a rope, for example)

208

it does not follow that it moves, a rejoinder that should not allow us to conclude that there are no causes for dogs moving or no sufficient conditions for making dogs move. Some supplementary premise is therefore required to make the argument plausible, but there is little to be gained seeking it. The conclusion of the argument appears to be false. For often we know how to get home and appear to be able to present people with directions which are sufficient for winning a race or a match; the *prima facie* case for our knowing that certain conditions are sufficient for certain achievements is strong. Of course, to know such conditions we need not know any general law or general theory for getting home or winning races in general, but that is another matter and is here irrelevant. I am not arguing that psychologists seek or ought to seek one general theory of perception which will explain each and every instance of or case of perception, but only that their program for seeking causal theories of perception is not misconceived, and further cannot be dismissed on the grounds that we never know that any statement of the form 'If S then T' is true when 'T' stands in place of an achievement term.

The counter argument against Mr. White is strengthened when we notice that he later agrees that "it is an analytically true statement that the word 'gaol' looks to normal persons in normal conditions as if it were the word 'gaol' "[7], a statement which can with suitable modifications be construed as stipulating a sufficient condition for seeing. But here Mr. White would argue that the statement he has presented is analytic and hence cannot serve as a premise in a causal explanation of perception.

He would further claim, I think, that it is as silly to claim that being a normal observer causes a man to see a blue object as blue, as it would be to insist that crossing a line first caused a man to win the race. We might, in both cases, explain why a given term applies, in the first "blue" to the object and in the second "winner of the race" to the man, but such explanations Mr. White would correctly aver are not causal ones.

Mr. White's reasoning though clear is not conclusive. Patently to give a causal explanation of why a man won a race or at least give good reasons for his winning, we need not restrict ourselves to the statement "he crossed the line first". Equally obvious is to explain why a man perceived we need not use the analytic statement "all normal observers, etc." Actually, 'A is normal' is not analytic and can provide us with an ingredient of a dispositional explanation of why the man saw what he did

209

on a given occasion. Further, since in giving a dispositional explanation we are tacitly assuming the realization of circumstances or conditions under which that disposition is made manifest, we may appeal to those conditions as the causally relevant factors and hence conclude that we have given not merely a dispositional but a causal explanation as well.

We have, however, only reached the half-way point. It will be contested that the causal explanation we have suggested differs in type from the causal explanations discussed to this point. We began discussing the availability of general laws which can serve as premises in causal explanations; we have, it will be contested, at best indicated the possibility of dispositional explanations which do not appeal to general laws. And since we have granted that 'all normal observers...' is analytic we will, it will be claimed in friendly rejoinder, be barred from finding suitable causal laws.

Notice, however, in our defense, that this rebuttal assumes, with Ryle, that no statement containing the name or description of an individual can be considered a causal law or general law; an assumption which I find dubious. Notice further that this rejoinder overlooks the methodological point that explanations of phenomena are frequently preceded by their description. 'Lethal drugs when taken in sufficient dosages cause death' is, I presume, as analytic as 'Normal observers in normal conditions see red things as red'. Yet it is clear that we can causally explain the death effects of lethal drugs by reducing them and their takers in physical-chemical terms. Similarly it may be analytic that a red object is seen as red, but it is not analytic that red objects have psychological characteristics Z, normal observers traits T, and hence not analytic that observers with traits T in normal circumstances see objects with characteristics Z as red.

Observe that we are not suggesting that the psychologist is to offer a redefinition or explication of 'does prove red', for that term is not tinkered with and appears both in the original statement and its replacement. Hamlyn's worries that psychologists may be incapable of offering proper explications of 'does perceive' can therefore be allayed. Neither are we definitely committed to the status of the schematic letter 'Z'. It is not part of the program that 'Z' stand in place of nonobservational predicates and hence the program does not enforce certain notorious versions of the causal theory of perception. However, everything said is compatible with another version of the causal theory, that which would insist that 'If A perceives X, then A senses T and T (or the sensings of T) is caused by X' (where 'X' stands in place of a name or a descriptive physical object) is analytic.

At all accounts it is this strategy of redirection which is not, I think, fully considered by Mr. White and others; and which I suggest was tacitly appealed to by Koffka when he raised the question "Why do things look the way they do".[7] And since such a strategy is not *prima facie* silly it will not do to argue with Hamlyn that the question "Why does X look the way it does when there is no expectation to the contrary is not to be considered a request for a scientific explanation, but something which has to do with epistemology."[8]

But not merely does Hamlyn overlook the point about redescription. In this and other sections of the book he confuses the psychological problem of explaining why and when we seek explanations with the logical one of assessing the legitimacy of the request for one. Normally, we do not ask why it snows in the winter and would ask for an explanation if it did begin snowing in August, but treatises on the weather are not exercises in epistemology. Similarly we do not ask why normal observers see red things as red but do ask why normal perceivers 'see' equal lines as unequal. But to be guided by our normal lack of curiosity and to conclude with Hamlyn that psychologists ought only to consider Muller-Lyer illusions and kindred phenomena is as silly as requesting physicists to close shop when they attempt to explain the normal behavior of magnets. Moreover, since Mr. Hamlyn does allow for causal explanations of illusions he cuts the ground from under one of his main arguments – to the effect that causal explanations of perception are impossible since "If our perceptions are caused there is no room to talk of correctness or error."[9]

Still Hamlyn's point might be saved if it were reinterpreted as a metalogical directive to the psychologist to begin building his theories on the assumption that normal perception is not to be explained; but only deviations from it are. But even if such a directive were adapted and proved fruitful it would be a mistake to conclude that normal perception cannot be explained; all that would follow is that normal perception is taken for granted in a given theory. Another theory might be invoked to explain normal perception.

211

Mr. Hamlyn still has his inning however. He seems to think that the mere fact that we have to appeal to skills dooms the attempt to get general laws, whether of perception or learning. For he says "the nature of skills is such that they can be performed under a variety of conditions. Something, however, may be said about the factors which tend to be sufficient to make people see things in a certain way".[10]

These are sober enough statements, but they raise more questions than they solve. I can decide at will whether to write this or that word; I cannot decide at will to see this or that except in the trivial sense that I can decide to place my body in a relevant position and open my eyes. To apply a general argument about skills to perception may therefore be the issue. Furthermore, though I know of no general laws for the exercise of intellectual skills, I presume that we can explain why Mr. Hamlyn exercised his skills and wrote what he did. Of course, here we can claim that Mr. Hamlyn had reasons for writing, and reasons are not causes. But unless I am mistaken we have no reasons for seeing, and hence the appropriateness of seeking causal laws and causal explanations of perception.

Columbia University, New York, N.Y.

REFERENCES

1. Gilbert Ryle, *Dilemmas*. London, 1956, p. 109–110.
2. Ryle, *l.c.* p. 110.
3. Ryle, *l.c.* p. 106.
4. D. W. Hamlyn, *The Psychology of Perception*. London, 1957, p. 92.
5. R. S. Peters, *The Concept of Motivation*. London, 1958, p. 14.
6. A. R. White, 'The Causal Theory of Perception', *Aristotelian Society* 25 (1961) 154.
7. White, *l.c.* p. 161.
8. Hamlyn, *l.c.* p. 48.
9. Hamlyn, *l.c.* p. 92.
10. Hamlyn, *l.c.* p. 114.